I0016899

Mastering SciPy

Implement state-of-the-art techniques to visualize solutions to challenging problems in scientific computing, with the use of the SciPy stack

Francisco J. Blanco-Silva

[PACKT] PUBLISHING

open source*
community experience distilled

BIRMINGHAM - MUMBAI

Mastering SciPy

Copyright © 2015 Packt Publishing

All rights reserved. No part of this book may be reproduced, stored in a retrieval system, or transmitted in any form or by any means, without the prior written permission of the publisher, except in the case of brief quotations embedded in critical articles or reviews.

Every effort has been made in the preparation of this book to ensure the accuracy of the information presented. However, the information contained in this book is sold without warranty, either express or implied. Neither the author, nor Packt Publishing, and its dealers and distributors will be held liable for any damages caused or alleged to be caused directly or indirectly by this book.

Packt Publishing has endeavored to provide trademark information about all of the companies and products mentioned in this book by the appropriate use of capitals. However, Packt Publishing cannot guarantee the accuracy of this information.

First published: November 2015

Production reference: 1301015

Published by Packt Publishing Ltd.
Livery Place
35 Livery Street
Birmingham B3 2PB, UK.

ISBN 978-1-78398-474-9

www.packtpub.com

Credits

Author
Francisco J. Blanco-Silva

Reviewers
Raiyan Kamal
Kristen Thyng
Patrick Varilly
Jonathan Whitmore

Commissioning Editor
Kartikey Pandey

Acquisition Editor
Shaon Basu

Content Development Editor
Nikhil Potdukhe

Technical Editor
Bharat Patil

Copy Editors
Tani Kothari
Merilyn Pereira

Project Coordinator
Judie Jose

Proofreader
Safis Editing

Indexer
Tejal Soni

Production Coordinator
Aparna Bhagat

Cover Work
Aparna Bhagat

About the Author

Francisco J. Blanco-Silva is the owner of a scientific consulting company called Tizona Scientific Solutions, a faculty member of the Department of Mathematics, and an associate member of the Interdisciplinary Mathematics Institute at the University of South Carolina. He obtained his formal training as an applied mathematician from Purdue University. He enjoys problem solving, learning, and teaching alike. Being an avid programmer and blogger, when it comes to writing, he relishes finding the common denominator among his passions and skills and making it available to everyone.

He wrote the prequel to this book, *Learning SciPy for Numerical and Scientific Computing*, *Packt Publishing*, and coauthored Chapter 5 of the book, *Modeling Nanoscale Imaging in Electron Microscopy*, *Springer*.

I will always be indebted to Bradley J. Lucier and Rodrigo Bañuelos for being constant sources of inspiration and for their guidance and teachings. Special thanks to my editors, Sriram Neelakantam, Bharat Patil, Nikhil Potdukhe, Mohammad Rizvi, and the many colleagues who have contributed by giving me encouragement and participating in helpful discussions. In particular, I would like to mention Parsa Bakhtary, Aaron Dutle, Edsel Peña, Pablo Sprechmann, Adam Taylor, and Holly Watson.

The most special thanks, without a doubt, goes to my wife and daughter. Grace's love and smiles alone provided all the motivation, enthusiasm, and skills to overcome the difficulties encountered during the writing of this book and everything that life threw at me ever since she was born.

About the Reviewers

Raiyan Kamal is a strong proponent of the open source movement and everything related to Python. He holds a bachelor's degree in computer science from BUET, Dhaka, Bangladesh, and a master's degree from the University of Windsor, Ontario, Canada. He has been working in the software industry for several years, developing software for mobile, web, and desktop platforms. Although he is in his early thirties, Raiyan feels that his boyhood has not ended yet. He often looks for hidden treasures in science, engineering, programming, art, and nature. He is currently working at IOU Concepts, exploring different ways of saying thank you. When he isn't on a computer, he plants trees and composts kitchen scraps.

Kristen Thyng has worked on scientific computing for most of her career. She has a bachelor's degree in physics from Whitman College, master's degree in applied mathematics from the University of Washington, and PhD in mechanical engineering from the University of Washington. She uses Python on a daily basis for analysis and visualization in physical oceanography at Texas A&M University, where she works as an assistant research scientist.

Jonathan Whitmore is a data scientist at Silicon Valley Data Science. He has a diverse range of interests and is excited by the challenges in data science and data engineering. Before moving into the tech industry, he worked as an astrophysicist in Melbourne, Australia, researching whether the fundamental physical constants have changed over the lifespan of the universe. He has a long-standing commitment to the public's understanding of science and technology, and has contributed to FOSS projects. He co-starred in the 3D IMAX film *Hidden Universe*, which was playing in theaters around the world at the time of writing this book. Jonathan is a sought-after conference speaker on science and technical topics. He received his PhD in physics from the University of California, San Diego, and graduated magna cum laude from Vanderbilt University with a bachelor's degree in science. He is also a triple major in physics (with honors), philosophy, and mathematics.

www.PacktPub.com

Support files, eBooks, discount offers, and more

For support files and downloads related to your book, please visit www.PacktPub.com.

Did you know that Packt offers eBook versions of every book published, with PDF and ePub files available? You can upgrade to the eBook version at www.PacktPub.com and as a print book customer, you are entitled to a discount on the eBook copy. Get in touch with us at service@packtpub.com for more details.

At www.PacktPub.com, you can also read a collection of free technical articles, sign up for a range of free newsletters and receive exclusive discounts and offers on Packt books and eBooks.

https://www2.packtpub.com/books/subscription/packtlib

Do you need instant solutions to your IT questions? PacktLib is Packt's online digital book library. Here, you can search, access, and read Packt's entire library of books.

Why subscribe?

- Fully searchable across every book published by Packt
- Copy and paste, print, and bookmark content
- On demand and accessible via a web browser

Free access for Packt account holders

If you have an account with Packt at www.PacktPub.com, you can use this to access PacktLib today and view 9 entirely free books. Simply use your login credentials for immediate access.

Table of Contents

Preface

The idea of writing *Mastering SciPy* arose but 2 months after publishing *Learning SciPy for Numerical and Scientific Computing*. During a presentation of that book at the University of South Carolina, I had the privilege of speaking about its contents to a heterogeneous audience of engineers, scientists, and students, each of them with very different research problems and their own set of preferred computational resources. In the weeks following that presentation, I helped a few professionals transition to a SciPy-based environment. During those sessions, we discussed how SciPy is, under the hood, the same set of algorithms (and often the same code) that they were already using. We experimented with some of their examples and systematically obtained comparable performance. We immediately saw the obvious benefit of a common environment based upon a robust scripting language. Through the SciPy stack, we discovered an easier way to communicate and share our results with colleagues, students, or employers. In all cases, the switch to the SciPy stack provided a faster setup for our groups, where newcomers could get up to speed quickly.

Everybody involved in the process went from *novice* to *advanced user*, and finally *mastered* the SciPy stack in no time. In most cases, the scientific background of the individuals with whom I worked made the transition seamless. The process toward mastering materialized when they were able to contrast the theory behind their research with the solutions offered. The *aha moment* always happened while replicating some of their experiments with a careful guidance and explanation of the process.

That is precisely the philosophy behind this book. I invite you to participate in similar sessions. Each chapter has been envisioned as a conversation with an individual with certain scientific needs expressed as numerical computations. Together, we discover relevant examples—the different possible ways to solve those problems, the theory behind them, and the pros and cons of each route.

The process of writing followed a similar path to obtain an engaging collection of examples. I entered into conversations with colleagues in several different fields. Each section clearly reflects these exchanges. This was crucial while engaged in the production of the most challenging chapters—the last four. To ensure the same quality throughout the book, always trying to commit to a rigorous set of standards, these chapters took much longer to be completed to satisfaction. Special mentions go to Aaron Dutle at NASA Langley Research Center, who helped shape parts of the chapter on computational geometry, and Parsa Bakhtary, a data analyst at Facebook, who inspired many of the techniques in the chapter on applications of statistical computing to data analysis.

It was an amazing journey that helped deepen my understanding of numerical methods, broadened my perspective in problem solving, and strengthened my scientific maturity. It is my wish that it has the same impact on you.

What this book covers

Chapter 1, Numerical Linear Algebra, presents an overview of the role of matrices to solve problems in scientific computing. It is a crucial chapter for understanding most of the processes and ideas of subsequent chapters. You will learn how to construct and store large matrices effectively in Python. We then proceed to reviewing basic manipulation and operations on them, followed by factorizations, solutions of matrix equations, and the computation of eigenvalues/eigenvectors.

Chapter 2, Interpolation and Approximation, develops advanced techniques to approximate functions, and their applications to scientific computing. This acts as a segway for the next two chapters.

Chapter 3, Differentiation and Integration, explores the different techniques to produce derivatives of functions and, more importantly, how to compute areas and volumes effectively by integration processes. This is the first of two chapters devoted to the core of numerical methods in scientific computing. This second part is also an introduction to *Chapter 5, Initial Value Problems for Ordinary Differential Equations* that mentions ordinary differential equations.

Chapter 4, Nonlinear Equations and Optimization, is a very technical chapter in which we discuss the best methods of obtaining the roots and extrema of systems of functions depending on the kinds of functions involved.

Chapter 5, Initial Value Problems for Ordinary Differential Equations, is the first of five chapters on applications to real-world problems. We show you, by example, the most popular techniques to solve systems of differential equations, as well as some applications.

Chapter 6, Computational Geometry, takes a tour of the most significant algorithms in this branch of computer science.

Chapter 7, Descriptive Statistics, is the first of two chapters on statistical computing and its applications to Data Analysis. In this chapter, we focus on probability and data exploration.

Chapter 8, Inference and Data Analysis, is the second chapter on Data Analysis. We focus on statistical inference, machine learning, and data mining.

Chapter 9, Mathematical Imaging, is the last chapter of this book. In it, we explore techniques for image compression, edition, restoration, and analysis.

What you need for this book

To work with the examples and try out the code of this book, all you need is a recent version of Python (2.7 or higher) with the SciPy stack: NumPy, the SciPy library, matplotlib, IPython, pandas, and SymPy. Although recipes to install all these independently are provided throughout the book, we recommend that you perform a global installation through a scientific Python distribution such as Anaconda.

Who this book is for

Although this book and technology are ultimately intended for applied mathematicians, engineers, and computer scientists, the material presented here is targeted at a broader audience. All that is needed is proficiency in Python, familiarity with iPython, some knowledge of the numerical methods in scientific computing, and a keen interest in developing serious applications in science, engineering, or data analysis.

Conventions

In this book, you will find a number of text styles that distinguish between different kinds of information. Here are some examples of these styles and an explanation of their meaning.

Code words in text, database table names, folder names, filenames, file extensions, pathnames, dummy URLs, user input, and Twitter handles are shown as follows: "We can include other contexts through the use of the include directive."

Any command-line input or output is written as follows:

```
In [7]: %time eigvals, v = spspla.eigsh(A, 5, which='SM')
CPU times: user 19.3 s, sys: 532 ms, total: 19.8 s
Wall time: 16.7 s
In [8]: print eigvals
[ 10.565523  10.663114  10.725135  10.752737  10.774503]
```

> Warnings or important notes appear in a box like this.

> Tips and tricks appear like this.

Reader feedback

Feedback from our readers is always welcome. Let us know what you think about this book—what you liked or disliked. Reader feedback is important for us as it helps us develop titles that you will really get the most out of.

To send us general feedback, simply e-mail feedback@packtpub.com, and mention the book's title in the subject of your message.

If there is a topic that you have expertise in and you are interested in either writing or contributing to a book, see our author guide at www.packtpub.com/authors.

Customer support

Now that you are the proud owner of a Packt book, we have a number of things to help you to get the most from your purchase.

Downloading the example code

You can download the example code files from your account at http://www.packtpub.com for all the Packt Publishing books you have purchased. If you purchased this book elsewhere, you can visit http://www.packtpub.com/support and register to have the files e-mailed directly to you. You can also download the code files from GitHub repository at https://github.com/blancosilva/Mastering-Scipy.

Downloading the color images of this book

We also provide you with a PDF file that has color images of the screenshots/ diagrams used in this book. The color images will help you better understand the changes in the output. You can download this file from `https://www.packtpub. com/sites/default/files/downloads/4749OS_ColorImages.pdf`.

Errata

Although we have taken every care to ensure the accuracy of our content, mistakes do happen. If you find a mistake in one of our books—maybe a mistake in the text or the code—we would be grateful if you could report this to us. By doing so, you can save other readers from frustration and help us improve subsequent versions of this book. If you find any errata, please report them by visiting `http://www.packtpub. com/submit-errata`, selecting your book, clicking on the **Errata Submission Form** link, and entering the details of your errata. Once your errata are verified, your submission will be accepted and the errata will be uploaded to our website or added to any list of existing errata under the Errata section of that title.

To view the previously submitted errata, go to `https://www.packtpub.com/books/ content/support` and enter the name of the book in the search field. The required information will appear under the **Errata** section.

Piracy

Piracy of copyrighted material on the Internet is an ongoing problem across all media. At Packt, we take the protection of our copyright and licenses very seriously. If you come across any illegal copies of our works in any form on the Internet, please provide us with the location address or website name immediately so that we can pursue a remedy.

Please contact us at `copyright@packtpub.com` with a link to the suspected pirated material.

We appreciate your help in protecting our authors and our ability to bring you valuable content.

Questions

If you have a problem with any aspect of this book, you can contact us at `questions@packtpub.com`, and we will do our best to address the problem.

1
Numerical Linear Algebra

The term **Numerical Linear Algebra** refers to the use of matrices to solve computational science problems. In this chapter, we start by learning how to construct these objects effectively in Python. We make an emphasis on importing large sparse matrices from repositories online. We then proceed to reviewing basic manipulation and operations on them. The next step is a study of the different matrix functions implemented in SciPy. We continue on to exploring different factorizations for the solution of matrix equations, and for the computation of eigenvalues and their corresponding eigenvectors.

Motivation

The following image shows a graph that represents a series of web pages (numbered from 1 to 8):

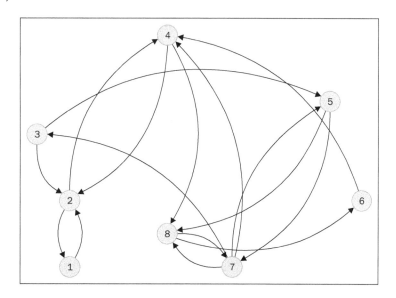

An arrow from a node to another indicates the existence of a link from the web page, represented by the sending node, to the page represented by the receiving node. For example, the arrow from node **2** to node **1** indicates that there is a link in web page **2** pointing to web page **1**. Notice how web page **4** has two outer links (to pages **2** and **8**), and there are three pages that link to web page **4** (pages **2**, **6**, and **7**). The pages represented by nodes **2**, **4**, and **8** seem to be the most popular at first sight.

Is there a mathematical way to actually express the popularity of a web page within a network? Researchers at Google came up with the idea of a `PageRank` to roughly estimate this concept by counting the number and quality of links to a page. It goes like this:

- We construct a transition matrix of this graph, `T={a[i,j]}`, in the following fashion: the entry `a[i,j]` is `1/k` if there is a link from web page `i` to web page `j`, and the total number of outer links in web page `i` amounts to `k`. Otherwise, the entry is just zero. The size of a transition matrix of N web pages is always $N \times N$. In our case, the matrix has size 8×8:

```
0    1/2  0    0    0    0    0    0
1    0    1/2  1/2  0    0    0    0
0    0    0    0    0    0    1/3  0
0    1/2  0    0    0    1    1/3  0
0    0    1/2  0    0    0    0    0
0    0    0    0    0    0    0    1/2
0    0    0    0    1/2  0    0    1/2
0    0    0    1/2  1/2  0    1/3  0
```

Let us open an iPython session and load this particular matrix to memory.

 Remember that in Python, indices start from zero, not one.

```
In [1]: import numpy as np, matplotlib.pyplot as plt, \
   ...: scipy.linalg as spla, scipy.sparse as spsp, \
   ...: scipy.sparse.linalg as spspla
In [2]: np.set_printoptions(suppress=True, precision=3)
In [3]: cols = np.array([0,1,1,2,2,3,3,4,4,5,6,6,6,7,7]); \
   ...: rows = np.array([1,0,3,1,4,1,7,6,7,3,2,3,7,5,6]); \
   ...: data = np.array([1., 0.5, 0.5, 0.5, 0.5, \
   ...:                  0.5, 0.5, 0.5, 0.5, 1., \
```

```
    ...:                     1./3, 1./3, 1./3, 0.5, 0.5])
In [4]: T = np.zeros((8,8)); \
    ...: T[rows,cols] = data
```

From the transition matrix, we create a `PageRank` matrix G by fixing a positive constant p between 0 and 1, and following the formula $G = (1-p)*T + p*B$ for a suitable damping factor p. Here, B is a matrix with the same size as T, with all its entries equal to `1/N`. For example, if we choose `p = 0.15`, we obtain the following `PageRank` matrix:

```
In [5]: G = (1-0.15) * T + 0.15/8; \
    ...: print G
[[ 0.019  0.444  0.019  0.019  0.019  0.019  0.019  0.019]
 [ 0.869  0.019  0.444  0.444  0.019  0.019  0.019  0.019]
 [ 0.019  0.019  0.019  0.019  0.019  0.019  0.302  0.019]
 [ 0.019  0.444  0.019  0.019  0.019  0.869  0.302  0.019]
 [ 0.019  0.019  0.444  0.019  0.019  0.019  0.019  0.019]
 [ 0.019  0.019  0.019  0.019  0.019  0.019  0.019  0.444]
 [ 0.019  0.019  0.019  0.019  0.444  0.019  0.019  0.444]
 [ 0.019  0.019  0.019  0.444  0.444  0.019  0.302  0.019]]
```

`PageRank` matrices have some interesting properties:

*   1 is an eigenvalue of multiplicity one.
*   1 is actually the largest eigenvalue; all the other eigenvalues are in modulus smaller than 1.
*   The eigenvector corresponding to eigenvalue 1 has all positive entries. In particular, for the eigenvalue 1, there exists a unique eigenvector with the sum of its entries equal to 1. This is what we call the `PageRank` vector.

A quick computation with `scipy.linalg.eig` finds that eigenvector for us:

```
In [6]: eigenvalues, eigenvectors = spla.eig(G); \
    ...: print eigenvalues
[ 1.000+0.j     -0.655+0.j     -0.333+0.313j -0.333-0.313j -0.171+0.372j
 -0.171-0.372j  0.544+0.j      0.268+0.j    ]
In [7]: PageRank = eigenvectors[:,0]; \
    ...: PageRank /= sum(PageRank); \
    ...: print PageRank.real
[ 0.117  0.232  0.048  0.219  0.039  0.086  0.102  0.157]
```

Those values correspond to the `PageRank` of each of the eight web pages depicted on the graph. As expected, the maximum value of those is associated to the second web page (`0.232`), closely followed by the fourth (`0.219`) and then the eighth web page (`0.157`). These values provide us with the information that we were seeking: the second web page is the most popular, followed by the fourth, and then, the eight.

> Note how this problem of networks of web pages has been translated into mathematical objects, to an equivalent problem involving matrices, eigenvalues, and eigenvectors, and has been solved with techniques of Linear Algebra.

The transition matrix is sparse: most of its entries are zeros. Sparse matrices with an extremely large size are of special importance in Numerical Linear Algebra, not only because they encode challenging scientific problems but also because it is extremely hard to manipulate them with basic algorithms.

Rather than storing to memory all values in the matrix, it makes sense to collect only the non-zero values instead, and use algorithms which exploit these smart storage schemes. The gain in memory management is obvious. These methods are usually faster for this kind of matrices and give less roundoff errors, since there are usually far less operations involved. This is another advantage of SciPy, since it contains numerous procedures to attack different problems where data is stored in this fashion. Let us observe its power with another example:

The University of Florida Sparse Matrix Collection is the largest database of matrices accessible online. As of January 2014, it contains 157 groups of matrices arising from all sorts of scientific disciplines. The sizes of the matrices range from very small (1×2) to insanely large (28 million $\times$ 28 million). More matrices are expected to be added constantly, as they arise in different engineering problems.

> More information about this database can be found in *ACM Transactions on Mathematical Software*, vol. 38, Issue 1, 2011, pp 1:1-1:25, by T.A. Davis and Y.Hu, or online at `http://www.cise.ufl.edu/research/sparse/matrices/`.

For example, the group with the most matrices in the database is the original Harwell-Boeing Collection, with 292 different sparse matrices. This group can also be accessed online at the Matrix Market: `http://math.nist.gov/MatrixMarket/`.

Each matrix in the database comes in three formats:

- **Matrix Market Exchange** format [Boisvert et al. 1997]
- **Rutherford-Boeing Exchange** format [Duff et al. 1997]
- **Proprietary Matlab** .mat format.

Let us import to our iPython session two matrices in the Matrix Market Exchange format from the collection, meant to be used in a solution of a least squares problem. These matrices are located at www.cise.ufl.edu/research/sparse/matrices/ Bydder/mri2.html.The numerical values correspond to phantom data acquired on a Sonata 1.5-T scanner (Siemens, Erlangen, Germany) using a **magnetic resonance imaging** (**MRI**) device. The object measured is a simulation of a human head made with several metallic objects. We download the corresponding tar bundle and untar it to get two ASCII files:

- mri2.mtx (the main matrix in the least squares problem)
- mri2_b.mtx (the right-hand side of the equation)

The first twenty lines of the file mri2.mtx read as follows:

```
%% MatrixMarket matrix coordinate real general
%-------------------------------------------------------------
% UF Sparse Matrix Collection, Tim Davis
% http://www.cise.ufl.edu/research/sparse/matrices/Bydder/mri2
% name: Bydder/mri2
% [MRI reconstruction (2), from Mark Bydder, UCSD]
% id: 1318
% date: 2005
% author: M. Bydder
% ed: T. Davis
% fields: title A name b id notes date author ed kind
% kind: computer graphics/vision problem
%-------------------------------------------------------------
% notes:
% x=lsqr(A,b); imagesc(abs(fftshift(fft2(reshape(x,384,384)))));
%-------------------------------------------------------------
63240 147456 569160
31992 1720 .053336731395584265
31992 1721 .15785917688901102
31992 1722 .07903055194318191
```

The first sixteen lines are comments, and give us some information about the generation of the matrix.

- The computer vision problem where it arose: An MRI reconstruction
- Author information: Mark Bydder, UCSD
- Procedures to apply to the data: Solve a least squares problem $A * x - b$, and posterior visualization of the result

The seventeenth line indicates the size of the matrix, 63240 rows × 147456 columns, as well as the number of non-zero entries in the data, 569160.

The rest of the file includes precisely 569160 lines, each containing two integer numbers, and a floating point number: These are the locations of the non-zero elements in the matrix, together with the corresponding values.

 We need to take into account that these files use the FORTRAN convention of starting arrays from 1, not from 0.

A good way to read this file into ndarray is by means of the function loadtxt in NumPy. We can then use scipy to transform the array into a sparse matrix with the function coo_matrix in the module scipy.sparse (coo stands for the coordinate internal format).

```
In [8]: rows, cols, data = np.loadtxt("mri2.mtx", skiprows=17, \
   ...:                               unpack=True)
In [9]: rows -= 1; cols -= 1;
In [10]: MRI2 = spsp.coo_matrix((data, (rows, cols)), \
   ....:                     shape=(63240,147456))
```

The best way to visualize the sparsity of this matrix is by means of the routine spy from the module matplotlib.pyplot.

```
In [11]: plt.spy(MRI2); \
   ....: plt.show()
```

Downloading the example code

You can download the example code files from your account at http://www.packtpub.com for all the Packt Publishing books you have purchased. If you purchased this book elsewhere, you can visit http://www.packtpub.com/support and register to have the files e-mailed directly to you.

We obtain the following image. Each pixel corresponds to an entry in the matrix; white indicates a zero value, and non-zero values are presented in different shades of blue, according to their magnitude (the higher, the darker):

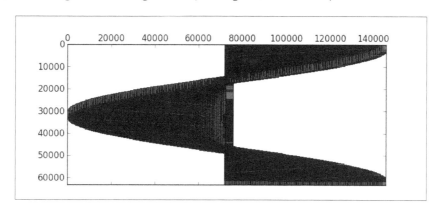

These are the first ten lines from the second file, `mri2_b.mtx`, which does not represent a sparse matrix, but a column vector:

```
%% MatrixMarket matrix array complex general
%-------------------------------------------------------------
% UF Sparse Matrix Collection, Tim Davis
% http://www.cise.ufl.edu/research/sparse/matrices/Bydder/mri2
% name: Bydder/mri2 : b matrix
%-------------------------------------------------------------
63240 1
-.07214859127998352 .037707749754190445
-.0729086771607399   .03763720765709877
-.07373382151126862 .03766685724258423
```

Those are six commented lines with information, one more line indicating the shape of the vector (63240 rows and 1 column), and the rest of the lines contain two columns of floating point values, the real and imaginary parts of the corresponding data. We proceed to read this vector to memory, solve the least squares problem suggested, and obtain the following reconstruction that represents a slice of the simulated human head:

```
In [12]: r_vals, i_vals = np.loadtxt("mri2_b.mtx", skiprows=7,
   ....:                              unpack=True)
In [13]: %time solution = spspla.lsqr(MRI2, r_vals + 1j*i_vals)
CPU times: user 4min 42s, sys: 1min 48s, total: 6min 30s
Wall time: 6min 30s
In [14]: from scipy.fftpack import fft2, fftshift
In [15]: img = solution[0].reshape(384,384); \
   ....: img = np.abs(fftshift(fft2(img)))
In [16]: plt.imshow(img); \
   ....: plt.show()
```

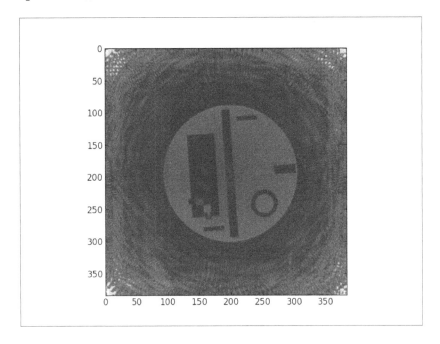

If interested in the theory behind the creation of this matrix and the particulars of this problem, read the article *On the optimality of the Gridding Reconstruction Algorithm*, by H. Sedarat and D. G. Nishimura, published in IEEE Trans. Medical Imaging, vol. 19, no. 4, pp. 306-317, 2000.

For matrices with a good structure, which are going to be exclusively involved in matrix multiplications, it is often possible to store the objects in smart ways. Let's consider an example.

A horizontal earthquake oscillation affects each floor of a tall building, depending on the natural frequencies of the oscillation of the floors. If we make certain assumptions, a model to quantize the oscillations on buildings with N floors can be obtained as a second-order system of N differential equations by competition: Newton's second law of force is set equal to the sum of Hooke's law of force, and the external force due to the earthquake wave.

These are the assumptions we will need:

- Each floor is considered a point of mass located at its center-of-mass. The floors have masses m[1], m[2], ..., m[N].

- Each floor is restored to its equilibrium position by a linear restoring force (Hooke's -k * elongation). The Hooke's constants for the floors are k[1], k[2], ..., k[N].

- The locations of masses representing the oscillation of the floors are x[1], x[2], ..., x[N]. We assume all of them functions of time and that at equilibrium, they are all equal to zero.

- For simplicity of exposition, we are going to assume no friction: all the damping effects on the floors will be ignored.

- The equations of a floor depend only on the neighboring floors.

Set M, the mass matrix, to be a diagonal matrix containing the floor masses on its diagonal. Set K, the Hooke's matrix, to be a tri-diagonal matrix with the following structure, for each row j, all the entries are zero except for the following ones:

- Column j-1, which we set to be k[j+1],
- Column j, which we set to -k[j+1]-k[j+1], and
- Column j+1, which we set to k[j+2].

Set H to be a column vector containing the external force on each floor due to the earthquake, and X, the column vector containing the functions x[j].

We have then the system: $M * X'' = K * X + H$. The homogeneous part of this system is the product of the inverse of M with K, which we denote as A.

To solve the homogeneous linear second-order system, $X'' = A * X$, we define the variable Y to contain $2*N$ entries: all N functions x[j], followed by their derivatives x'[j]. Any solution of this second-order linear system has a corresponding solution on the first-order linear system $Y' = C * Y$, where C is a block matrix of size 2*N × 2*N. This matrix C is composed by a block of size $N × N$ containing only zeros, followed horizontally by the identity (of size $N × N$), and below these two, the matrix A followed horizontally by another N × N block of zeros.

It is not necessary to store this matrix C into memory, or any of its factors or blocks. Instead, we will make use of its structure, and use a linear operator to represent it. Minimal data is then needed to generate this operator (only the values of the masses and the Hooke's coefficients), much less than any matrix representation of it.

Let us show a concrete example with six floors. We indicate first their masses and Hooke's constants, and then, proceed to construct a representation of A as a linear operator:

```
In [17]: m = np.array([56., 56., 56., 54., 54., 53.]); \
    ....: k = np.array([561., 562., 560., 541., 542., 530.])
In [18]: def Axv(v):
    ....:     global k, m
    ....:     w = v.copy()
    ....:     w[0] = (k[1]*v[1] - (k[0]+k[1])*v[0])/m[0]
    ....:     for j in range(1, len(v)-1):
    ....:         w[j] = k[j]*v[j-1] + k[j+1]*v[j+1] - \
    ....:                 (k[j]+k[j+1])*v[j]
    ....:         w[j] /= m[j]
    ....:     w[-1] = k[-1]*(v[-2]-v[-1])/m[-1]
    ....:     return w
    ....:
In [19]: A = spspla.LinearOperator((6,6), matvec=Axv, matmat=Axv,
    ....:                           dtype=np.float64)
```

The construction of C is very simple now (much simpler than that of its matrix!):

```
In [20]: def Cxv(v):
   ....:         n = len(v)/2
   ....:         w = v.copy()
   ....:         w[:n] = v[n:]
   ....:         w[n:] = A * v[:n]
   ....:         return w
   ....:
In [21]: C = spspla.LinearOperator((12,12), matvec=Cxv, matmat=Cxv,
   ....:                                   dtype=np.float64)
```

A solution of this homogeneous system comes in the form of an action of the exponential of C: $Y(t) = expm(C*t)* Y(0)$, where expm() here denotes a matrix exponential function. In SciPy, this operation is performed with the routine expm_multiply in the module scipy.sparse.linalg.

For example, in our case, given the initial value containing the values x[1](0)=0, ..., x[N](0)=0, x'[1](0)=1, ..., x'[N](0)=1, if we require a solution Y(t) for values of t between 0 and 1 in steps of size 0.1, we could issue the following:

It has been reported in some installations that, in the next step, a matrix for C must be given instead of the actual linear operator (thus contradicting the manual). If this is the case in your system, simply change C in the next lines to its matrix representation.

```
In [22]: initial_condition = np.zeros(12); \
   ....: initial_condition[6:] = 1
In [23]: Y = spspla.exp_multiply(C, np.zeros(12), start=0,
   ....:                                   stop=1, num=10)
```

The oscillations of the six floors during the first second can then be calculated and plotted. For instance, to view the oscillation of the first floor, we could issue the following:

```
In [24]: plt.plot(np.linspace(0,1,10), Y[:,0]); \
   ....: plt.xlabel('time (in seconds)'); \
   ....: plt.ylabel('oscillation')
```

We obtain the following plot. Note how the first floor rises in the first tenth of a second, only to drop from 0.1 to 0.9 seconds from its original height to almost under a meter and then, start a slow rise:

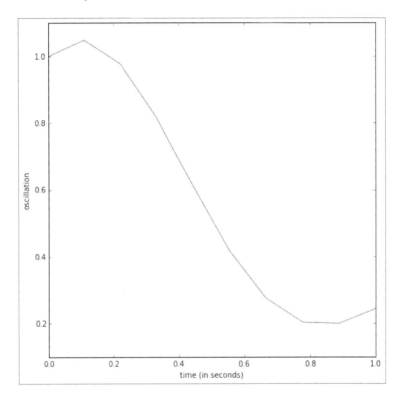

For more details about systems of differential equations, and how to solve them with actions of exponentials, read, for example, the excellent book, *Elementary Differential Equations 10 ed.*, by William E. Boyce and Richard C. DiPrima. Wiley, 2012.

These three examples illustrate the goal of this first chapter, Numerical Linear Algebra. In Python, this is accomplished first by storing the data in a matrix form, or as a related linear operator, by means of any of the following classes:

- `numpy.ndarray` (making sure that they are two-dimensional)
- `numpy.matrix`
- `scipy.sparse.bsr_matrix` (Block Sparse Row matrix)
- `scipy.sparse.coo_matrix` (Sparse Matrix in **COO**rdinate format)
- `scipy.sparse.csc_matrix` (Compressed Sparse Column matrix)

- `scipy.sparse.csr_matrix` (Compressed Sparse Row matrix)
- `scipy.sparse.dia_matrix` (Sparse matrix with **DIA**gonal storage)
- `scipy.sparse.dok_matrix` (Sparse matrix based on a Dictionary of Keys)
- `scipy.sparse.lil_matrix` (Sparse matrix based on a linked list)
- `scipy.sparse.linalg.LinearOperator`

As we have seen in the examples, the choice of different classes obeys mainly to the sparsity of data and the algorithms that we are to apply to them.

 We will learn when to apply these choices in the following sections.

This choice then dictates the modules that we use for the different algorithms: `scipy.linalg` for generic matrices and both `scipy.sparse` and `scipy.sparse.linalg` for sparse matrices or linear operators. These three SciPy modules are compiled on top of the highly optimized computer libraries BLAS (written in Fortran77), LAPACK (in Fortran90), ARPACK (in Fortran77), and SuperLU (in C).

 For a better understanding of these underlying packages, read the description and documentation from their creators:

- **BLAS**: `netlib.org/blas/faq.html`
- **LAPACK**: `netlib.org/lapack/lapack-3.2.html`
- **ARPACK**: `www.caam.rice.edu/software/ARPACK/`
- **SuperLU**: `crd-legacy.lbl.gov/~xiaoye/SuperLU/`

Most of the routines in these three SciPy modules are wrappers to functions in the mentioned libraries. If we so desire, we also have the possibility to call the underlying functions directly. In the `scipy.linalg` module, we have the following:

- `scipy.linalg.get_blas_funcs` to call routines from BLAS
- `scipy.linalg.get_lapack_funcs` to call routines from LAPACK

For example, if we want to use the BLAS function NRM2 to compute Frobenius norms:

```
In [25]: blas_norm = spla.get_blas_func('nrm2')
In [26]: blas_norm(np.float32([1e20]))
Out[26]: 1.0000000200408773e+20
```

Creation of matrices and linear operators

In the first part of this chapter, we are going to focus on the effective creation of matrices. We start by recalling some different ways to construct a basic matrix as an `ndarray` instance class, including an enumeration of all the special matrices already included in NumPy and SciPy. We proceed to examine the possibilities of constructing complex matrices from basic ones. We review the same concepts within the `matrix` instance class. Next, we explore in detail the different ways to input sparse matrices. We finish the section with the construction of linear operators.

 We assume familiarity with `ndarray` creation in NumPy, as well as data types (dtype), indexing, routines for the combination of two or more arrays, array manipulation, or extracting information from these objects. In this chapter, we will focus on the functions, methods, and routines that are significant to matrices alone. We will disregard operations if their outputs have no translation into linear algebra equivalents. For a primer on `ndarray`, we recommend you to browse through *Chapter 2, Top-Level SciPy* of *Learning SciPy for Numerical and Scientific Computing, Second Edition*. For a quick review of Linear Algebra, we recommend Hoffman and Kunze, *Linear Algebra 2nd Edition, Pearson*, 1971.

Constructing matrices in the ndarray class

We may create matrices from data as `ndarray` instances in three different ways: manually from standard input, by assigning to each entry a value from a function, or by retrieving the data from external files.

Constructor	Description
`numpy.array(object)`	Create a matrix from `object`
`numpy.diag(arr, k)`	Create diagonal matrix with entries of array `arr` on diagonal `k`
`numpy.fromfunction(function, shape)`	Create a matrix by executing a function over each coordinate
`numpy.fromfile(fname)`	Create a matrix from a text or binary file (basic)
`numpy.loadtxt(fname)`	Create a matrix from a text file (advanced)

Let us create some example matrices to illustrate some of the functions defined in the previous table. As before, we start an iPython session:

```
In [1]: import numpy as np, matplotlib.pyplot as plt, \
   ...: scipy.linalg as spla, scipy.sparse as spsp, \
   ...: scipy.sparse.linalg as spspla
In [2]: A = np.array([[1,2],[4,16]]); \
   ...: A
Out[2]:
array([[ 1,   2],
       [ 4,  16]])
In [3]: B = np.fromfunction(lambda i,j: (i-1)*(j+1),
   ...:                     (3,2), dtype=int); \
   ...: print B
   ...:
 [[-1 -2]
  [ 0   0]
  [ 1   2]]
In [4]: np.diag((1j,4))
Out[4]:
array([[ 0.+1.j,   0.+0.j],
       [ 0.+0.j,   4.+0.j]])
```

Special matrices with predetermined zeros and ones can be constructed with the following functions:

Constructor	Description
numpy.empty(shape)	Array of a given shape, entries not initialized
numpy.eye(N, M, k)	2-D array with ones on the k-th diagonal, and zeros elsewhere
numpy.identity(n)	Identity array
numpy.ones(shape)	Array with all entries equal to one
numpy.zeros(shape)	Array with all entries equal to zero
numpy.tri(N, M, k)	Array with ones at and below the given diagonal, zeros otherwise

All these constructions, except `numpy.tri`, have a companion function `xxx_like` that creates `ndarray` with the requested characteristics and with the same shape and data type as another source `ndarray` class:

```
In [5]: np.empty_like(A)
Out[5]:
array([[140567774850560, 140567774850560],
       [     4411734640, 562954363882576]])
```

Of notable importance are arrays constructed as numerical ranges.

Constructor	Description
`numpy.arange(stop)`	Evenly spaced values within an interval
`numpy.linspace(start, stop)`	Evenly spaced numbers over an interval
`numpy.logspace(start, stop)`	Evenly spaced numbers on a log scale
`numpy.meshgrid`	Coordinate matrices from two or more coordinate vectors
`numpy.mgrid`	`nd_grid` instance returning dense multi-dimensional `meshgrid`
`numpy.ogrid`	`nd_grid` instance returning open multi-dimensional `meshgrid`

Special matrices with numerous applications in linear algebra can be easily called from within NumPy and the module `scipy.linalg`.

Constructor	Description
`scipy.linalg.circulant(arr)`	Circulant matrix generated by 1-D array `arr`
`scipy.linalg.companion(arr)`	Companion matrix of polynomial with coefficients coded by `arr`
`scipy.linalg.hadamard(n)`	Sylvester's construction of a Hadamard matrix of size $n \times n$. n must be a power of 2
`scipy.linalg.hankel(arr1, arr2)`	Hankel matrix with `arr1` as the first column and `arr2` as the last column
`scipy.linalg.hilbert(n)`	Hilbert matrix of size $n \times n$
`scipy.linalg.invhilbert(n)`	The inverse of a Hilbert matrix of size $n \times n$
`scipy.linalg.leslie(arr1, arr2)`	Leslie matrix with fecundity array `arr1` and survival coefficients `arr2`
`scipy.linalg.pascal(n)`	$n \times n$ truncations of the Pascal matrix of binomial coefficients

Constructor	Description
`scipy.linalg.toeplitz(arr1, arr2)`	Toeplitz array with first column `arr1` and first row `arr2`
`numpy.vander(arr)`	Van der Monde matrix of array `arr`

For instance, one fast way to obtain all binomial coefficients of orders up to a large number (the corresponding Pascal triangle) is by means of a precise Pascal matrix. The following example shows how to compute these coefficients up to order 13:

```
In [6]: print spla.pascal(13, kind='lower')
```

```
[[  1   0   0   0   0   0   0   0   0   0   0   0   0]
 [  1   1   0   0   0   0   0   0   0   0   0   0   0]
 [  1   2   1   0   0   0   0   0   0   0   0   0   0]
 [  1   3   3   1   0   0   0   0   0   0   0   0   0]
 [  1   4   6   4   1   0   0   0   0   0   0   0   0]
 [  1   5  10  10   5   1   0   0   0   0   0   0   0]
 [  1   6  15  20  15   6   1   0   0   0   0   0   0]
 [  1   7  21  35  35  21   7   1   0   0   0   0   0]
 [  1   8  28  56  70  56  28   8   1   0   0   0   0]
 [  1   9  36  84 126 126  84  36   9   1   0   0   0]
 [  1  10  45 120 210 252 210 120  45  10   1   0   0]
 [  1  11  55 165 330 462 462 330 165  55  11   1   0]
 [  1  12  66 220 495 792 924 792 495 220  66  12   1]]
```

Besides these basic constructors, we can always stack arrays in different ways:

Constructor	Description
`numpy.concatenate((A1, A2, ...))`	Join matrices together
`numpy.hstack((A1, A2, ...))`	Stack matrices horizontally
`numpy.vstack((A1, A2, ...))`	Stack matrices vertically
`numpy.tile(A, reps)`	Repeat a matrix a certain number of times (given by `reps`)
`scipy.linalg.block_diag(A1,A2, ...)`	Create a block diagonal array

Let us observe some of these constructors in action:

```
In [7]: np.tile(A, (2,3))    # 2 rows, 3 columns
Out[7]:
array([[ 1,   2,   1,   2,   1,   2],
       [ 4,  16,   4,  16,   4,  16],
       [ 1,   2,   1,   2,   1,   2],
       [ 4,  16,   4,  16,   4,  16]])
In [8]: spla.block_diag(A,B)
Out[8]:
array([[ 1,   2,   0,   0],
       [ 4,  16,   0,   0],
       [ 0,   0,  -1,  -2],
       [ 0,   0,   0,   0],
       [ 0,   0,   1,   2]])
```

Constructing matrices in the matrix class

For the `matrix` class, the usual way to create a matrix directly is to invoke either `numpy.mat` or `numpy.matrix`. Observe how much more comfortable is the syntax of `numpy.matrix` than that of `numpy.array`, in the creation of a matrix similar to A. With this syntax, different values separated by commas belong to the same row of the matrix. A semi-colon indicates a change of row. Notice the casting to the `matrix` class too!

```
In [9]: C = np.matrix('1,2;4,16'); \
   ...: C
Out[9]:
matrix([[ 1,   2],
        [ 4,  16]])
```

These two functions also transform any `ndarray` into `matrix`. There is a third function that accomplishes this task: `numpy.asmatrix`:

```
In [10]: np.asmatrix(A)
Out[10]:
matrix([[ 1,   2],
        [ 4,  16]])
```

For arrangements of matrices composed by blocks, besides the common stack operations for `ndarray` described before, we have the extremely convenient function `numpy.bmat`. Note the similarity with the syntax of `numpy.matrix`, particularly the use of commas to signify horizontal concatenation and semi-colons to signify vertical concatenation:

```
In [11]: np.bmat('A;B')         In [12]: np.bmat('A,C;C,A')
Out[11]:                        Out[12]:
matrix([[ 1,  2],               matrix([[ 1,  2,  1,  2],
        [ 4, 16],                       [ 4, 16,  4, 16],
        [-1, -2],                       [ 1,  2,  1,  2],
        [ 0,  0],                       [ 4, 16,  4, 16]])
        [ 1,  2]])
```

Constructing sparse matrices

There are seven different ways to input sparse matrices. Each format is designed to make a specific problem or operation more efficient. Let us go over them in detail:

Method	Name	Optimal use
BSR	Block Sparse Row	Efficient arithmetic, provided the matrix contains blocks.
COO	Coordinate	Fast and efficient construction format. Efficient methods to convert to the CSC and CSR formats.
CSC	Compressed Sparse Column	Efficient matrix arithmetic and column slicing. Relatively fast matrix-vector product.
CSR	Compressed Sparse Row	Efficient matrix arithmetic and row slicing. Fastest to perform matrix-vector products.
DIA	Diagonal storage	Efficient for construction and storage if the matrix contains long diagonals of non-zero entries.
DOK	Dictionary of keys	Efficient incremental construction and access of individual matrix entries.
LIL	Row-based linked list	Flexible slicing. Efficient for changes to matrix sparsity.

They can be populated in up to five ways, three of which are common to every sparse matrix format:

- They can cast to sparse any generic matrix. The `lil` format is the most effective with this method:

```
In [13]: A_coo = spsp.coo_matrix(A); \
   ....: A_lil = spsp.lil_matrix(A)
```

- They can cast to a specific sparse format another sparse matrix in another sparse format:

```
In [14]: A_csr = spsp.csr_matrix(A_coo)
```

- Empty sparse matrices of any shape can be constructed by indicating the shape and `dtype`:

```
In [15]: M_bsr = spsp.bsr_matrix((100,100), dtype=int)
```

They all have several different extra input methods, each specific to their storage format.

- **Fancy indexing**: As we would do with any generic matrix. This is only possible with the LIL or DOK formats:

```
In [16]: M_lil = spsp.lil_matrix((100,100), dtype=int)

In [17]: M_lil[25:75, 25:75] = 1

In [18]: M_bsr[25:75, 25:75] = 1

NotImplementedError    Traceback (most recent call last)
<ipython-input-18-d9fa1001cab8> in <module>()
----> 1 M_bsr[25:75, 25:75] = 1

[...]/scipy/sparse/bsr.pyc in __setitem__(self, key, val)
    297
    298    def __setitem__(self,key,val):
--> 299        raise NotImplementedError
    300
    301    #####################

NotImplementedError:
```

- **Dictionary of keys**: This input system is most effective when we create, update, or search each element one at a time. It is efficient only for the LIL and DOK formats:

```
In [19]: M_dok = spsp.dok_matrix((100,100), dtype=int)

In [20]: position = lambda i, j: ((i<j) & ((i+j)%10==0))

In [21]: for i in range(100):
```

```
....:        for j in range(100):
....:            M_dok[i,j] = position(i,j)
....:
```

- **Data, rows, and columns**: This is common to four formats: BSR, COO, CSC, and CSR. This is the method of choice to import sparse matrices from the **Matrix Market Exchange** format, as illustrated at the beginning of the chapter.

> With the data, rows, and columns input method, it is a good idea to always include the option `shape` in the construction. In case this is not provided, the size of the matrix will be inferred from the largest coordinates from the rows and columns, resulting possibly in a matrix of a smaller size than required.

- **Data, indices, and pointers**: This is common to three formats: BSR, CSC, and CSR. It is the method of choice to import sparse matrices from the Rutherford-Boeing Exchange format.

> The Rutherford-Boeing Exchange format is an updated version of the Harwell-Boeing format. It stores the matrix as three vectors: `pointers_v`, `indices_v`, and `data`. The row indices of the entries of the *j*th column are located in positions `pointers_v(j)` through `pointers_v(j+1)-1` of the vector `indices_v`. The corresponding values of the matrix are located at the same positions, in the vector data.

Let us show by example how to read an interesting matrix in the Rutherford-Boeing matrix exchange format, `Pajek/football`. This 35 × 35 matrix with 118 non-zero entries can be found in the collection at `www.cise.ufl.edu/research/sparse/matrices/Pajek/football.html`.

It is an adjacency matrix for a network of all the national football teams that attended the FIFA World Cup celebrated in France in 1998. Each node in the network represents one country (or national football team) and the links show which country exported players to another country.

This is a printout of the `football.rb` file:

```
Pajek/football; 1998; L. Krempel; ed: V. Batagelj                    |1474
             12                2           5              5
iua                           35          35            118              0
(20I4)            (26I3)             (26I3)
   1    3    6    8    8   10   12   12   12   12   28   28   46   59   71   74   74   74   90   90
  94   94   94   95   95  101  101  105  105  105  107  108  115  118  118  119
   7    9    8   16   35   24   29    9   28   24   35    2    3    4    8   11   16   17   22   24   25   26   29   31   33   34
  35    1    2    4    5    8    9   10   11   16   18   22   24   25   26   28   29   34   35    2    3    5    8   10   18   21
  22   24   30   31   34   35    2   11   16   18   19   25   26   29   30   33   34   35    8   26   29    1    2    3    5    7
   8    9   10   11   16   22   24   25   26   34   35    5    8   21   34   28    3   11   24   29   33   35    4    5    8   22
  11   26   22    4    8   11   16   24   29   35    7    9   24   24
   1    3    1    1    1    2    1    2   10    1    1    7    2    4    1    3    2    3    3    1    1    3    2    2    2    1
   2    4    1    1    4    2    1    1    1    4    2    4    4    5    1    2    6    9    2    1    2    1    7    1    1    1
   1    3    1    3    1    1    1    6    2    2    7    4   12    2    7    2    1    4    1    2    2    9    1    2    5    1
   2    1    2    1    4    1    2    2    1    7    1    1    2    4    2    1    2    1    3    1    1    2    1    1    2    3
   3    3    1    4    2    1    1    3    4    3    1    2    1    1
```

The header of the file (the first four lines) contains important information:

- The first line provides us with the title of the matrix, `Pajek/football; 1998; L. Krempel; ed: V. Batagelj,` and a numerical key for identification purposes `MTRXID=1474`.

- The second line contains four integer values: `TOTCRD=12` (lines containing significant data after the header; see In [24]), `PTRCRD=2` (number of lines containing pointer data), `INDCRD=5` (number of lines containing indices data), and `VALCRD=2` (number of lines containing the non-zero values of the matrix). Note that it must be $TOTCRD = PTRCRD + INDCRD + VALCRD$.

- The third line indicates the matrix type `MXTYPE=(iua)`, which in this case stands for an integer matrix, unsymmetrical, compressed column form. It also indicates the number of rows and columns (`NROW=35`, `NCOL=35`), and the number of non-zero entries (`NNZERO=118`). The last entry is not used in the case of a compressed column form, and it is usually set to zero.

- The fourth column contains the Fortran formats for the data in the following columns. `PTRFMT=(20I4)` for the pointers, `INDFMT=(26I3)` for the indices, and `VALFMT=(26I3)` for the non-zero values.

We proceed to opening the file for reading, storing each line after the header in a Python list, and extracting from the relevant lines of the file, the data we require to populate the vectors `indptr`, `indices`, and `data`. We finish by creating the corresponding sparse matrix called `football` in the CSR format, with the `data`, `indices`, `pointers` method:

```
In [22]: f = open("football.rb", 'r'); \
   ....: football_list = list(f); \
   ....: f.close()
In [23]: football_data = np.array([])
In [24]: for line in range(4, 4+12):
   ....:        newdata = np.fromstring(football_list[line], sep=" ")
   ....:        football_data = np.append(football_data, newdata)
   ....:
In [25]: indptr = football_data[:35+1] - 1; \
   ....: indices = football_data[35+1:35+1+118] - 1; \
   ....: data = football_data[35+1+118:]
In [26]: football = spsp.csr_matrix((data, indices, indptr),
   ....:                                shape=(35,35))
```

At this point, it is possible to visualize the network with its associated graph, with the help of a Python module called `networkx`. We obtain the following diagram depicting as nodes the different countries. Each arrow between the nodes indicates the fact that the originating country has exported players to the receiving country:

 `networkx` is a Python module to deal with complex networks. For more information, visit their Github project pages at `networkx.github.io`.

One way to accomplish this task is as follows:

```
In [27]: import networkx
In [28]: G = networkx.DiGraph(football)
In [29]: f = open("football_nodename.txt"); \
   ....: m = list(f); \
   ....: f.close()
In [30]: def rename(x): return m[x]
In [31]: G = networkx.relabel_nodes(G, rename)
```

```
In [32]: pos = networkx.spring_layout(G)
In [33]: networkx.draw_networkx(G, pos, alpha=0.2, node_color='w',
   ....:                        edge_color='b')
```

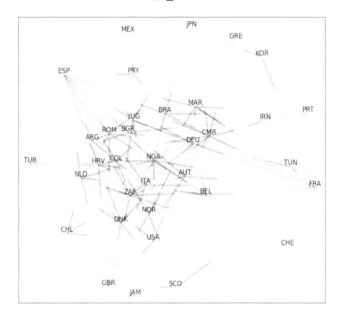

The module `scipy.sparse` borrows from NumPy some interesting concepts to create constructors and special matrices:

Constructor	Description
`scipy.sparse.diags(diagonals, offsets)`	Sparse matrix from diagonals
`scipy.sparse.rand(m, n, density)`	Random sparse matrix of prescribed density
`scipy.sparse.eye(m)`	Sparse matrix with ones in the main diagonal
`scipy.sparse.identity(n)`	Identity sparse matrix of size $n \times n$

Both functions `diags` and `rand` deserve examples to show their syntax. We will start with a sparse matrix of size 14×14 with two diagonals: the main diagonal contains 1s, and the diagonal below contains 2s. We also create a random matrix with the function `scipy.sparse.rand`. This matrix has size 5×5, with 25 percent non-zero elements (`density=0.25`), and is crafted in the LIL format:

```
In [34]: diagonals = [[1]*14, [2]*13]
In [35]: print spsp.diags(diagonals, [0,-1]).todense()
```

```
[[ 1.  0.  0.  0.  0.  0.  0.  0.  0.  0.  0.  0.  0.  0.]
 [ 2.  1.  0.  0.  0.  0.  0.  0.  0.  0.  0.  0.  0.  0.]
 [ 0.  2.  1.  0.  0.  0.  0.  0.  0.  0.  0.  0.  0.  0.]
 [ 0.  0.  2.  1.  0.  0.  0.  0.  0.  0.  0.  0.  0.  0.]
 [ 0.  0.  0.  2.  1.  0.  0.  0.  0.  0.  0.  0.  0.  0.]
 [ 0.  0.  0.  0.  2.  1.  0.  0.  0.  0.  0.  0.  0.  0.]
 [ 0.  0.  0.  0.  0.  2.  1.  0.  0.  0.  0.  0.  0.  0.]
 [ 0.  0.  0.  0.  0.  0.  2.  1.  0.  0.  0.  0.  0.  0.]
 [ 0.  0.  0.  0.  0.  0.  0.  2.  1.  0.  0.  0.  0.  0.]
 [ 0.  0.  0.  0.  0.  0.  0.  0.  2.  1.  0.  0.  0.  0.]
 [ 0.  0.  0.  0.  0.  0.  0.  0.  0.  2.  1.  0.  0.  0.]
 [ 0.  0.  0.  0.  0.  0.  0.  0.  0.  0.  2.  1.  0.  0.]
 [ 0.  0.  0.  0.  0.  0.  0.  0.  0.  0.  0.  2.  1.  0.]
 [ 0.  0.  0.  0.  0.  0.  0.  0.  0.  0.  0.  0.  2.  1.]]
In [36]: S_25_lil = spsp.rand(5, 5, density=0.25, format='lil')
In [37]: S_25_lil
Out[37]:
<5x5 sparse matrix of type '<type 'numpy.float64'>'
        with 6 stored elements in LInked List format>
In [38]: print S_25_lil
  (0, 0)      0.186663044982
  (1, 0)      0.127636181284
  (1, 4)      0.918284870518
  (3, 2)      0.458768884701
  (3, 3)      0.533573291684
  (4, 3)      0.908751420065
In [39]: print S_25_lil.todense()
[[ 0.18666304  0.          0.          0.          0.        ]
 [ 0.12763618  0.          0.          0.          0.91828487]
 [ 0.          0.          0.          0.          0.        ]
 [ 0.          0.          0.45876888  0.53357329  0.        ]
 [ 0.          0.          0.          0.90875142  0.        ]]
```

Similar to the way we combined `ndarray` instances, we have some clever ways to combine sparse matrices to construct more complex objects:

Constructor	Description
`scipy.sparse.bmat(blocks)`	Sparse matrix from sparse sub-blocks
`scipy.sparse.hstack(blocks)`	Stack sparse matrices horizontally
`scipy.sparse.vstack(blocks)`	Stack sparse matrices vertically

Linear operators

A linear operator is basically a function that takes as input a column vector and outputs another column vector, by left multiplication of the input with a matrix. Although technically, we could represent these objects just by handling the corresponding matrix, there are better ways to do this.

Constructor	Description
`scipy.sparse.linalg.` `LinearOperator(shape, matvec)`	Common interface for performing matrix vector products
`scipy.sparse.linalg.aslinearoperator(A)`	Return A as `LinearOperator`

In the `scipy.sparse.linalg` module, we have a common interface that handles these objects: the `LinearOperator` class. This class has only the following two attributes and three methods:

- `shape`: The shape of the representing matrix
- `dtype`: The data type of the matrix
- `matvec`: To perform multiplication of a matrix with a vector
- `rmatvec`: To perform multiplication by the conjugate transpose of a matrix with a vector
- `matmat`: To perform multiplication of a matrix with another matrix

Its usage is best explained through an example. Consider two functions that take vectors of size 3, and output vectors of size 4, by left multiplication with two respective matrices of size 4 × 3. We could very well define these functions with lambda predicates:

```
In [40]: H1 = np.matrix("1,3,5; 2,4,6; 6,4,2; 5,3,1"); \
    ....: H2 = np.matrix("1,2,3; 1,3,2; 2,1,3; 2,3,1")
In [41]: L1 = lambda x: H1.dot(x); \
```

```
    ....: L2 = lambda x: H2.dot(x)
In [42]: print L1(np.ones(3))
[[  9.  12.  12.   9.]]
In [43]: print L2(np.tri(3,3))
  [[ 6.   5.   3.]
   [ 6.   5.   2.]
   [ 6.   4.   3.]
   [ 6.   4.   1.]]
```

Now, one issue arises when we try to add/subtract these two functions, or multiply any of them by a scalar. Technically, it should be as easy as adding/subtracting the corresponding matrices, or multiplying them by any number, and then performing the required left multiplication again. But that is not the case.

For instance, we would like to write `(L1+L2)(v)` instead of `L1(v) + L2(v)`. Unfortunately, doing so will raise an error:

```
TypeError: unsupported operand type(s) for +: 'function' and
'function'
```

Instead, we may instantiate the corresponding linear operators and manipulate them at will, as follows:

```
In [44]: Lo1 = spspla.aslinearoperator(H1); \
    ....: Lo2 = spspla.aslinearoperator(H2)
In [45]: Lo1 - 6 * Lo2
Out[45]: <4x3 _SumLinearOperator with dtype=float64>
In [46]: print Lo1 * np.ones(3)
[  9.  12.  12.   9.]
In [47]: print (Lo1-6*Lo2) * np.tri(3,3)
[[-27. -22. -13.]
 [-24. -20.  -6.]
 [-24. -18. -16.]
 [-27. -20.  -5.]]
```

Linear operators are a great advantage when the amount of information needed to describe the product with the related matrix is less than the amount of memory needed to store the non-zero elements of the matrix.

For instance, a permutation matrix is a square binary matrix (ones and zeros) that has exactly one entry in each row and each column. Consider a large permutation matrix, say 1024×1024, formed by four blocks of size 512×512: a zero block followed horizontally by an identity block, on top of an identity block followed horizontally by another zero block. We may store this matrix in three different ways:

```
In [47]: P_sparse = spsp.diags([[1]*512, [1]*512], [512,-512], \
    ....:                          dtype=int)
In [48]: P_dense = P_sparse.todense()
In [49]: mv = lambda v: np.roll(v, len(v)/2)
In [50]: P_lo = spspla.LinearOperator((1024,1024), matvec=mv, \
    ....:                          matmat=mv, dtype=int)
```

In the sparse case, `P_sparse`, we may think of this as the storage of just 1024 integer numbers. In the dense case, `P_dense`, we are technically storing 1048576 integer values. In the case of the linear operator, it actually looks like we are not storing anything! The function `mv` that indicates how to perform the multiplications has a much smaller footprint than any of the related matrices. This is also reflected in the time of execution of the multiplications with these objects:

```
In [51]: %timeit P_sparse * np.ones(1024)
10000 loops, best of 3: 29.7 µs per loop
In [52]: %timeit P_dense.dot(np.ones(1024))
100 loops, best of 3: 6.07 ms per loop
In [53]: %timeit P_lo * np.ones(1024)
10000 loops, best of 3: 25.4 µs per loop
```

Basic matrix manipulation

The emphasis of the second part of this chapter is on mastering the following operations:

- Scalar multiplication, matrix addition, and matrix multiplication
- Traces and determinants
- Transposes and inverses
- Norms and condition numbers

Scalar multiplication, matrix addition, and matrix multiplication

Let us start with the matrices stored with the `ndarray` class. We accomplish scalar multiplication with the `*` operator, and the matrix addition with the `+` operator. But for matrix multiplication we will need the instance method `dot()` or the `numpy.dot` function, since the operator `*` is reserved for element-wise multiplication:

```
In [54]: 2*A
Out[54]:
array([[ 2,  4],
       [ 8, 32]])
In [55]: A + 2*A
Out[55]:
array([[ 3,  6],
       [12, 48]])
```

```
In [56]: A.dot(2*A)        In [56]: np.dot(A, 2*A)
Out[56]:                   Out[56]:
array([[ 18,  68],         array([[ 18,  68],
       [136, 528]])               [136, 528]])
In [57]: A.dot(B)
ValueError: objects are not aligned
In [58]: B.dot(A)          In [58]: np.dot(B, A)
Out[58]:                   Out[58]:
array([[ -9, -34],         array([[ -9, -34],
       [  0,   0],                [  0,   0],
       [  9,  34]])               [  9,  34]])
```

The matrix class makes matrix multiplication more intuitive: the operator `*` can be used instead of the `dot()` method. Note also how matrix multiplication between different instance classes `ndarray` and a matrix is always casted to a `matrix` instance class:

```
In [59]: C * B
ValueError: shapes (2,2) and (3,2) not aligned: 2 (dim 1) != 3 (dim 0)
In [60]: B * C
Out[60]:
matrix([[ -9, -34],
        [  0,   0],
        [  9,  34]])
```

For sparse matrices, both scalar multiplication and addition work well with the obvious operators, even if the two sparse classes are not the same. Note the resulting class casting after each operation:

```
In [61]: S_10_coo = spsp.rand(5, 5, density=0.1, format='coo')
In [62]: S_25_lil + S_10_coo
Out[62]: <5x5 sparse matrix of type '<type 'numpy.float64'>'
        with 8 stored elements in Compressed Sparse Row format>
In [63]: S_25_lil * S_10_coo
Out[63]: <5x5 sparse matrix of type '<type 'numpy.float64'>'
        with 4 stored elements in Compressed Sparse Row format>
```

 numpy.dot does not work well for matrix multiplication of a sparse matrix with a generic. We must use the operator * instead.

```
In [64]: S_100_coo = spsp.rand(2, 2, density=1, format='coo')
In [65]: np.dot(A, S_100_coo)
Out[66]:
array([[ <2x2 sparse matrix of type '<type 'numpy.float64'>'
  with 4 stored elements in COOrdinate format>,
        <2x2 sparse matrix of type '<type 'numpy.float64'>'
  with 4 stored elements in COOrdinate format>],
       [ <2x2 sparse matrix of type '<type 'numpy.float64'>'
  with 4 stored elements in COOrdinate format>,
        <2x2 sparse matrix of type '<type 'numpy.float64'>'
  with 4 stored elements in COOrdinate format>]], dtype=object)
In [67]: A * S_100_coo
Out[68]:
array([[  1.81 ,    1.555],
       [ 11.438,   11.105]])
```

Traces and determinants

The traces of a matrix are the sums of the elements on the diagonals (assuming always increasing indices in both dimensions). For generic matrices, we compute them with the instance method trace(), or with the function numpy.trace:

```
In [69]: A.trace()          In [71]: C.trace()
Out[69]: 17                 Out[71]: matrix([[17]])
In [70]: B.trace()          In [72]: np.trace(B, offset=-1)
Out[70]: -1                 Out[72]: 2
```

In order to compute the determinant of generic square matrices, we need the function `det` in the module `scipy.linalg`:

```
In [73]: spla.det(C)
Out[73]: 8.0
```

Transposes and inverses

Transposes can be computed with any of the two instance methods `transpose()` or `T`, for any of the two classes of generic matrices:

```
In [74]: B.transpose()        In [75]: C.T
Out[74]:                      Out[75]:
array([[-1,  0,  1],          matrix([[ 1,  4],
       [-2,  0,  2]])                 [ 2, 16]])
```

Hermitian transpose can be computed for the `matrix` class with the instance method `H`:

```
In [76]: D = C * np.diag((1j,4)); print D      In [77]: print D.H
[[  0.+1.j    8.+0.j]                          [[  0.-1.j    0.-4.j]
 [  0.+4.j   64.+0.j]]                           [  8.-0.j   64.-0.j]]
```

Inverses of non-singular square matrices are computed for the `ndarray` class with the function `inv` in the module `scipy.linalg`. For the `matrix` class, we may also use the instance method `I`. For non-singular square sparse matrices, we may use the function `inv` in the module `scipy.sparse.linalg`.

Inverses of sparse matrices are seldom sparse. For this reason, it is not recommended to perform this operation with the `scipy.sparse.inv` function. One possible way to go around this issue is to convert the matrix to generic with the `todense()` instance method, and use `scipy.linear.inv` instead.

But due to the difficulty of inverting large matrices, it is often beneficial to compute approximations to the inverse, instead. The function `spilu` in the module `scipy.sparse.linalg` provides us with a very fast algorithm to perform this computation for square sparse matrices in CSC format. This algorithm is based on *LU decompositions*, and coded internally as a wrapper of a function from the library `SuperLU`. Its use is rather complex, and we are going to postpone its study until we explore `matrix` factorizations.

```
In [78]: E = spsp.rand(512, 512, density=1).todense()
In [79]: S_100_csc = spsp.rand(512, 512, density=1, format='csc')
```

```
In [80]: %timeit E.I
10 loops, best of 3: 28.7 ms per loop
In [81]: %timeit spspla.inv(S_100_csc)
1 loops, best of 3: 1.99 s per loop
```

> In the execution of sparse inverses, if the input matrix is not in the CSC or CSR format, we will get a warning:
>
> ```
> /scipy/sparse/linalg/dsolve/linsolve.py:88:
> SparseEfficiencyWarning: spsolve requires A be CSC or CSR
> matrix format
>   warn('spsolve requires A be CSC or CSR matrix format',
> SparseEfficiencyWarning)
> /scipy/sparse/linalg/dsolve/linsolve.py:103:
> SparseEfficiencyWarning: solve requires b be CSC or CSR
> matrix format
> ```

The Moore-Penrose pseudo-inverse can be computed for any kind of matrix (not necessarily square) with either routines the pinv or the pinv2 in the module scipy. linalg. The first method, pinv, resorts to solving a least squares problem to compute the pseudo-inverse. The function pinv2 computes the pseudo-inverse by a method based on singular value decompositions. For Hermitian matrices, or matrices that are symmetric with no complex coefficients, we also have a third function called pinvh, which is based on eigenvalue decompositions.

It is known that in the case of square non-singular matrices, the inverse and pseudo-inverse are the same. This simple example shows the times of computation of the inverses of a large generic symmetric matrix with the five methods described:

```
In [82]: F = E + E.T     # F is symmetric
In [83]: %timeit F.I
1 loops, best of 3: 24 ms per loop
In [84]: %timeit spla.inv(F)
10 loops, best of 3: 28 ms per loop
In [85]: %timeit spla.pinvh(E)
1 loops, best of 3: 120 ms per loop
In [86]: %timeit spla.pinv2(E)
1 loops, best of 3: 252 ms per loop
In [87]: %timeit spla.pinv(F)
1 loops, best of 3: 2.21 s per loop
```

Norms and condition numbers

For generic matrices, we have seven different standard norms in `scipy.linalg`. We can summarize them in the following table:

Constructor	Description
norm(A,numpy.inf)	Sum of absolute values of entries in each row. Pick the largest value.
norm(A,-numpy.inf)	Sum of absolute values of entries in each row. Pick the smallest value.
norm(A,1)	Sum of absolute values of entries in each column. Pick the largest value.
norm(A,-1)	Sum of absolute values of entries in each column. Pick the smallest value.
norm(A,2)	Largest eigenvalue of the matrix.
norm(A,-2)	Smallest eigenvalue of the matrix.
norm(A,'fro') or norm(A,'f')	Frobenius norm: the square root of the trace of the product $A.H * A$.

```
In [88]: [spla.norm(A,s) for s in (np.inf,-np.inf,-1,1,-2,2,'fro')]
Out[88]: [20, 3, 5, 18, 0.48087417361008861, 16.636368595013604,
16.643316977093239]
```

For sparse matrices, we can always compute norms by applying the `todense()` instance method prior to computation. But when the sizes of the matrices are too large, this is very impractical. In those cases, the best we can get for the 1-norm is a lower bound, thanks to the function `onenormest` in the module `scipy.sparse.linalg`:

```
In [89]: spla.norm(S_100_csc.todense(), 1) - \
   ....: spspla.onenormest(S_100_csc)
Out[89]: 0.0
```

As for the 2-norms, we may find the values of the smallest and the largest eigenvalue, but only for square matrices. We have two algorithms in the module `scipy.sparse.linalg` that perform this task: eigs (for generic square matrices) and eigsh for real symmetric matrices. We will explore them in detail when we discuss matrix decompositions and factorizations in the next section.

Note the subtle difference between the norm computations from SciPy and NumPy. For example, in the case of the Frobenius norm, `scipy.linalg.norm` is based directly on the BLAS function called NRM2, while `numpy.linalg.norm` is equivalent to a purely straightforward computation of the form `sqrt(add.reduce((x.conj() * x).real))`. The advantage of the code based on BLAS, besides being much faster, is clear when some of the data is too large or too small in single-precision arithmetic. This is shown in the following example:

```
In [89]: a = np.float64([1e20]); \
   ....: b = np.float32([1e20])
In [90]: [np.linalg.norm(a), spla.norm(a)]
Out[90]: [1e+20, 1e+20]
In [91]: np.linalg.norm(b)
[...]/numpy/linalg/linalg.py:2056: RuntimeWarning: overflow encountered
in multiply
  return sqrt(add.reduce((x.conj() * x).real, axis=None))
Out[91]: inf
In [92]: spla.norm(b)
Out[92]: 1.0000000200408773e+20
```

This brings us inevitably to a discussion about the computation of the condition number of a non-singular square matrix A. This value measures how much the output of the solution to the linear equation $A * x = b$ will change when we make small changes to the input argument b. If this value is close to one, we can rest assured that the solution is going to change very little (we say then that the system is well-conditioned). If the condition number is large, we know that there might be issues with the computed solutions of the system (and we say then that it is ill-conditioned).

The computation of this condition number is performed by multiplying the norm of A with the norm of its inverse. Note that there are different condition numbers, depending on the norm that we choose for the computation. These values can also be computed for each of the pre-defined norms with the function `numpy.linalg.cond`, although we need to be aware of its obvious limitations.

```
In [93]: np.linalg.cond(C, -np.inf)
Out[93]: 1.875
```

Matrix functions

A matrix function is a function that maps a square matrix to another square matrix via a power series. These should not be confused with vectorization: the application of any given function of one variable to each element of a matrix. For example, it is not the same to compute the square of a square matrix, `A.dot(A)` (for example, `In [8]`), than a matrix with all the elements of `A` squared (examples `In [5]` through `In []`).

 To make the proper distinction in notation, we will write `A^2` to denote the actual square of a square matrix and `A^n` to represent the subsequent powers (for all positive integers *n*).

Constructor	Description
`scipy.linalg.funm(A, func, disp)`	Extension of a scalar-valued function called `func` to a matrix
`scipy.linalg.fractional_matrix_power(A, t)`	Fractional matrix power
`scipy.linalg.expm(A)` or `scipy.sparse.linalg.expm(A)`	Matrix exponential
`scipy.sparse.linalg.expm_multiply(A,B)`	Action of the matrix exponential of `A` on `B`
`scipy.linalg.expm_frechet(A, E)`	Frechet derivative of the matrix exponential in the `E` direction
`scipy.linalg.cosm(A)`	Matrix cosine
`scipy.linalg.sinm(A)`	Matrix sine
`scipy.linalg.tanm(A)`	Matrix tangent
`scipy.linalg.coshm(A)`	Hyperbolic matrix cosine
`scipy.linalg.sinhm(A)`	Hyperbolic matrix sine
`scipy.linalg.tanhm(A)`	Hyperbolic matrix tangent
`scipy.linalg.signm(A)`	Matrix sign function
`scipy.linalg.sqrtm(A, disp, blocksize)`	Matrix square root
`scipy.linalg.logm(A, disp)`	Matrix logarithm

```
In [1]: import numpy as np, scipy as sp; \
   ...: import scipy.linalg as spla
In [2]: np.set_printoptions(suppress=True, precision=3)
In [3]: def square(x): return x**2
In [4]: A = spla.hilbert(4); print A
```

```
[[ 1.       0.5      0.333    0.25 ]
 [ 0.5     0.333    0.25     0.2  ]
 [ 0.333   0.25     0.2      0.167]
 [ 0.25    0.2      0.167    0.143]]
In [5]: print square(A)
[[ 1.       0.25     0.111    0.062]
 [ 0.5     0.333    0.25     0.2  ]
 [ 0.333   0.25     0.2      0.167]
 [ 0.25    0.2      0.167    0.143]]
In [6]: print A*A
[[ 1.       0.25     0.111    0.062]
 [ 0.25    0.111    0.062    0.04 ]
 [ 0.111   0.062    0.04     0.028]
 [ 0.062   0.04     0.028    0.02 ]]
In [7]: print A**2
[[ 1.       0.25     0.111    0.062]
 [ 0.25    0.111    0.062    0.04 ]
 [ 0.111   0.062    0.04     0.028]
 [ 0.062   0.04     0.028    0.02 ]]
In [8]: print A.dot(A)
[[ 1.424   0.8      0.567    0.441]
 [ 0.8     0.464    0.333    0.262]
 [ 0.567   0.333    0.241    0.19 ]
 [ 0.441   0.262    0.19     0.151]]
```

The actual powers A^n of a matrix is the starting point for the definition of any matrix function. In the module numpy.linalg we have the routine matrix_power to perform this operation. We can also achieve this result with the generic function funm or with the function fractional_matrix_power, both of them in the module scipy.linalg.

```
In [9]: print np.linalg.matrix_power(A, 2)
[[ 1.424   0.8      0.567    0.441]
 [ 0.8     0.464    0.333    0.262]
 [ 0.567   0.333    0.241    0.19 ]
 [ 0.441   0.262    0.19     0.151]]
In [10]: print spla.fractional_matrix_power(A, 2)
[[ 1.424   0.8      0.567    0.441]
```

```
 [ 0.8     0.464   0.333   0.262]
 [ 0.567   0.333   0.241   0.19 ]
 [ 0.441   0.262   0.19    0.151]]
In [11]: print spla.funm(A, square)
[[ 1.424   0.8     0.567   0.441]
 [ 0.8     0.464   0.333   0.262]
 [ 0.567   0.333   0.241   0.19 ]
 [ 0.441   0.262   0.19    0.151]]
```

To compute any matrix function, theoretically, we first express the function as a power series, by means of its Taylor expansion. Then, we apply the input matrix into an approximation to that expansion (since it is impossible to add matrices ad infinitum). Most matrix functions necessarily carry an error of computation, for this reason. In the `scipy.linalg` module, the matrix functions are coded following this principle.

- Note that there are three functions with an optional Boolean parameter `disp`. To understand the usage of this parameter, we must remember that most matrix functions compute approximations, with an error of computation. The parameter `disp` is set to `True` by default, and it produces a warning if the error of approximation is large. If we set `disp` to `False`, instead of a warning we will obtain the 1-norm of the estimated error.

- The algorithms behind the functions `expm`, the action of an exponential over a matrix, `expm_multiply`, and the Frechet derivative of an exponential, `expm_frechet`, use Pade approximations instead of Taylor expansions. This allows for more robust and accurate calculations. All the trigonometric and hyperbolic trigonometric functions base their algorithm in easy computations involving `expm`.

- The generic matrix function called `funm` and the square-root function called `sqrtm` apply clever algorithms that play with the *Schur decomposition* of the input matrix, and proper algebraic manipulations with the corresponding eigenvalues. They are still prone to roundoff errors but are much faster and more accurate than any algorithm based on Taylor expansions.

- The matrix sign function called `signm` is initially an application of `funm` with the appropriate function, but should this approach fail, the algorithm takes a different approach based on iterations that converges to a decent approximation to the solution.

- The functions `logm` and `fractional_matrix_power` (when the latter is applied to non-integer powers) use a very complex combination (and improvement!) of Pade approximations and Schur decompositions.

We will explore Schur decompositions when we deal with matrix factorizations related to eigenvalues. In the meantime, if you are interested in learning the particulars of these clever algorithms, read their descriptions in Golub and Van Loan, *Matrix Computations 4 edition*, Johns Hopkins Studies in the Mathematical Sciences, vol. 3.

For details on the improvements to Schur-Pade algorithms, as well as the algorithm behind Frechet derivatives of the exponential, refer to:

- Nicholas J. Higham and Lijing Lin *An Improved Schur-Pade Algorithm for Fractional Powers of a Matrix and Their Frechet Derivatives*
- Awad H. Al-Mohy and Nicholas J. Higham *Improved Inverse Scaling and Squaring Algorithms for the Matrix Logarithm*, in SIAM Journal on Scientific Computing, 34 (4)

Matrix factorizations related to solving matrix equations

The concept of matrix decompositions is what makes Numerical Linear Algebra an efficient tool in Scientific Computing. If the matrix representing a problem is simple enough, any basic generic algorithm can find the solutions optimally (that is, fast, with minimal storage of data, and without a significant roundoff error). But, in real life, this situation seldom occurs. What we do in the general case is finding a suitable matrix factorization and tailoring an algorithm that is optimal on each factor, thus gaining on each step an obvious advantage. In this section, we explore the different factorizations included in the modules `scipy.linalg` and `scipy.sparse.linalg` that help us achieve a robust solution to matrix equations.

Relevant factorizations

We have the following factorizations in this category:

Pivoted LU decomposition

It is always possible to perform a factorization of a square matrix A as a product $A = P \cdot L \cdot U$ of a permutation matrix P (which performs a permutation of the rows of A), a lower triangular matrix L, and an upper triangular matrix U:

Constructor	Description
`scipy.linalg.lu(A)`	Pivoted LU decomposition
`scipy.linalg.lu_factor(A)`	Pivoted LU decomposition
`scipy.sparse.linalg.splu(A)`	Pivoted LU decomposition
`scipy.sparse.linalg.spilu(A)`	Incomplete pivoted LU decomposition

Cholesky decomposition

For a square, symmetric, and positive definite matrix A, we can realize the matrix as the product $A = U^T \cdot U$ of an upper triangular matrix U with its transpose, or as the product $A = L^T \cdot L$ of a lower triangular matrix L with its transpose. All the diagonal entries of U or L are strictly positive numbers:

Constructor	Description
`scipy.linalg.cholesky(A)`	Cholesky decomposition
`scipy.linalg.cholesky_banded(AB)`	Cholesky decomposition for Hermitian positive-definite banded matrices

QR decomposition

We can realize any matrix of size m × n as the product $A=Q \cdot R$ of a square orthogonal matrix Q of size m × m, with an upper triangular matrix R of the same size as A.

Constructor	Description
`scipy.linalg.qr(A)`	QR decomposition of a matrix

Singular value decomposition

We can realize any matrix A as the product $A = U \cdot D \cdot V^H$ of a unitary matrix U with a diagonal matrix D (where all entries in the diagonal are positive numbers), and the Hermitian transpose of another unitary matrix V. The values on the diagonal of D are called the singular values of A.

Constructor	Description
`scipy.linalg.svd(A)`	Singular value decomposition
`scipy.linalg.svdvals(A)`	Singular values
`scipy.linalg.diagsvd(s, m, n)`	Diagonal matrix of an SVD, from singular values `s` and prescribed size
`scipy.sparse.linalg.svds(A)`	Largest k singular values/vectors of a sparse matrix

Matrix equations

In SciPy, we have robust algorithms to solve any matrix equation based on the following cases:

- Given a square matrix A, and a right-hand side b (which can be a one-dimensional vector or another matrix with the same number of rows as A), the basic systems are as follows:
  - $A \cdot x = b$
  - $A^T \cdot x = b$
  - $A^H \cdot x = b$

- Given any matrix A (not necessarily square) and a right-hand side vector/matrix b of an appropriate size, the least squares solution to the equation $A \cdot x = b$. This is, finding a vector x that minimizes the Frobenius norm of the expression $A \cdot x - b$.

- For the same case as before, and an extra damping coefficient d, the regularized least squares solution to the equation $A \cdot x = b$ that minimizes the functional `norm(A * x - b, 'f')**2 + d**2 * norm(x, 'f')**2`.

- Given square matrices A and B, and a right-hand side matrix Q with appropriate sizes, the Sylvester system is $A \cdot X + X \cdot B = Q$.

- For a square matrix A and matrix Q of an appropriate size, the continuous Lyapunov equation is $A \cdot X + X \cdot A^H = Q$.

- For matrices A and Q, as in the previous case, the discrete Lyapunov equation is $X - A \cdot X \cdot A^H = Q$.

- Given square matrices A, Q, and R, and another matrix B with an appropriate size, the continuous algebraic Riccati equation is $A^T \cdot X + X \cdot A - X \cdot B \cdot R^{-1} \cdot B^T \cdot X + Q = 0$.

- For matrices as in the previous case, the Discrete Algebraic Riccati equation is $X = A^T \cdot X \cdot A - (A^T \cdot X \cdot B) \cdot (R + B^T \cdot X \cdot B)^{-1} \cdot (B^T \cdot X \cdot A) + Q$.

In any case, mastering matrix equations with SciPy basically means identifying the matrices involved and choosing the most adequate algorithm in the libraries to perform the requested operations. Besides being able to compute a solution with the least possible amount of roundoff error, we need to do so in the fastest possible way, and by using as few memory resources as possible.

Back and forward substitution

Let us start with the easiest possible case: The basic system of linear equations $A \cdot x = b$ (or the other two variants), where A is a generic lower or upper triangular square matrix. In theory, these systems are easily solved by forward substitution (for lower triangular matrices) or back substitution (for upper triangular matrices). In SciPy, we accomplish this task with the function `solve_triangular` in the module `scipy.linalg`.

For this initial example, we will construct A as a lower triangular Pascal matrix of size 1024 × 1024, where the non-zero values have been filtered: odd values are turned into ones, while even values are turned into zeros. The right-hand side b is a vector with 1024 ones.

```
In [1]: import numpy as np, \
   ...: scipy.linalg as spla, scipy.sparse as spsp, \
   ...: scipy.sparse.linalg as spspla
In [2]: A = (spla.pascal(1024, kind='lower')%2 != 0)
In [3]: %timeit spla.solve_triangular(A, np.ones(1024))
10 loops, best of 3: 6.64 ms per loop
```

To solve the other related systems that involve the matrix A, we employ the optional parameter `trans` (by default set to 0 or N, giving the basic system $A \cdot x = b$). If `trans` is set to T or 1, we solve the system $A^T \cdot x = b$ instead. If `trans` is set to C or 2, we solve $A^H \cdot x = b$ instead.

The function `solve_triangular` is a wrapper for the LAPACK function `trtrs`.

Basic systems: banded matrices

The next cases in terms of algorithm simplicity are those of basic systems $A \cdot x = b$, where A is a square banded matrix. We use the routines `solve_banded` (for a generic banded matrix) or `solveh_banded` (for a generic real symmetric of complex Hermitian banded matrix). Both of them belong to the module `scipy.linalg`.

The functions `solve_banded` and `solveh_banded` are wrappers for the LAPACK functions GBSV, and PBSV, respectively.

Neither function accepts a matrix in the usual format. For example, since `solveh_`
`banded` expects a symmetric banded matrix, the function requires as input only the
elements of the diagonals on and under/over the main diagonal, stored sequentially
from the top to the bottom.

This input method is best explained through a concrete example. Take the following
symmetric banded matrix:

```
 2 -1  0  0  0  0
-1  2 -1  0  0  0
 0 -1  2 -1  0  0
 0  0 -1  2 -1  0
 0  0  0 -1  2 -1
 0  0  0  0 -1  2
```

The size of the matrix is 6×6, and there are only three non-zero diagonals, two of
which are identical due to symmetry. We collect the two relevant non-zero diagonals
in `ndarray` of size 2×6 in one of two ways, as follows:

- If we decide to input the entries from the upper triangular matrix, we collect
  first the diagonals from the top to the bottom (ending in the main diagonal),
  right justified:

  ```
  *  -1 -1 -1 -1 -1
  2   2  2  2  2  2
  ```

- If we decide to input the entries from the lower triangular matrix, we collect
  the diagonals from the top to the bottom (starting from the main diagonal),
  left justified:

  ```
   2  2  2  2  2  2
  -1 -1 -1 -1 -1  *
  In [4]: B_banded = np.zeros((2,6)); \
     ...: B_banded[0,1:] = -1; \
     ...: B_banded[1,:] = 2
  In [5]: spla.solveh_banded(B_banded, np.ones(6))
  Out[5]: array([ 3.,   5.,   6.,   6.,   5.,   3.])
  ```

For a non-symmetric banded square matrix, we use `solve_banded` instead, and the input matrix also needs to be stored in this special way:

- Count the number of non-zero diagonals under the main diagonal (set that to l). Count the number of non-zero diagonals over the main diagonal (set that to u). Set `r = l + u + 1`.

- If the matrix has size $n \times n$, create `ndarray` with n columns and r rows. We refer to this storage as a matrix in the AB form, or an AB matrix, for short.

- Store in the AB matrix only the relevant non-zero diagonals, from the top to the bottom, in order. Diagonals over the main diagonal are right justified; diagonals under the main diagonal are left justified.

Let us illustrate this process with another example. We input the following matrix:

```
 2 -1  0  0  0  0
-1  2 -1  0  0  0
 3 -1  2 -1  0  0
 0  3 -1  2 -1  0
 0  0  3 -1  2 -1
 0  0  0  3 -1  2
In [6]: C_banded = np.zeros((4,6)); \
   ...: C_banded[0,1:] = -1; \
   ...: C_banded[1,:] = 2; \
   ...: C_banded[2,:-1] = -1; \
   ...: C_banded[3,:-2] = 3; \
   ...: print C_banded
[[ 0. -1. -1. -1. -1. -1.]
 [ 2.  2.  2.  2.  2.  2.]
 [-1. -1. -1. -1. -1.  0.]
 [ 3.  3.  3.  3.  0.  0.]]
```

To call the solver, we need to input manually the number of diagonals over and under the diagonal, together with the AB matrix and the right-hand side of the system:

```
In [7]: spla.solve_banded((2,1), C_banded, np.ones(6))
Out[7]:
array([ 0.86842105,  0.73684211, -0.39473684,  0.07894737,
        1.76315789,  1.26315789])
```

Let us examine the optional parameters that we can include in the call of these two functions:

Parameter	Default values	Description
l_and_u	(int, int)	Number of non-zero lower/upper diagonals
ab	Matrix in AB format	A banded square matrix
b	ndarray	Right-hand side
overwrite_ab	Boolean	Discard data in ab
overwrite_b	Boolean	Discard data in b
check_finite	Boolean	Whether to check that input matrices contain finite numbers

All the functions in the scipy.linalg module that require matrices as input and output either a solution to a system of equations, or a factorization, have two optional parameters with which we need to familiarize: overwrite_x (for each matrix/vector in the input) and check_finite. They are both Boolean.

The overwrite options are set to False by default. If we do not care about retaining the values of the input matrices, we may use the same object in the memory to perform operations, rather than creating another object with the same size in the memory. We gain speed and use fewer resources in such a case.

The check_finite option is set to True by default. In the algorithms where it is present, there are optional checks for the integrity of the data. If at any given moment, any of the values is (+/-)numpy.inf or NaN, the process is halted, and an exception is raised. We may turn this option off, thus resulting in much faster solutions, but the code might crash if the data is corrupted at any point in the computations.

The function solveh_banded has an extra optional Boolean parameter, lower, which is initially set to False. If set to True, we must input the lower triangular matrix of the target AB matrix instead of the upper one (with the same input convention as before).

Basic systems: generic square matrices

For solutions of basic systems where A is a generic square matrix, it is a good idea to factorize A so that some (or all) of the factors are triangular and then apply back and forward substitution, where appropriate. This is the idea behind pivoted LU and Cholesky decompositions.

If matrix A is real symmetric (or complex Hermitian) and positive definite, the optimal strategy goes through applying any of the two possible Cholesky decompositions $A = U^H \cdot U$ or $A = L \cdot L^H$ with the U and L upper/lower triangular matrices.

For example, if we use the form with the upper triangular matrices, the solution of the basic system of equations $A \cdot x = b$ turns into $U^H \cdot U \cdot x = b$. Set $y = U \cdot x$ and solve the system $U^H \cdot y = b$ for y by forward substitution. We have now a new triangular system $U \cdot x = y$ that we solve for x, by back substitution.

To perform the solution of such a system with this technique, we first compute the factorization by using either the functions `cholesky`, `cho_factor` or `cholesky_banded`. The output is then used in the solver `cho_solve`.

For Cholesky decompositions, the three relevant functions called `cholesky`, `cho_factor`, and `cholesky_banded` have a set of options similar to those of `solveh_banded`. They admit an extra Boolean option lower (set by default to `False`) that decides whether to output a lower or an upper triangular factorization. The function `cholesky_banded` requires a matrix in the AB format as input.

Let us now test the Cholesky decomposition of matrix B with all three methods:

```
In [8]: B = spsp.diags([[-1]*5, [2]*6, [-1]*5], [-1,0,1]).todense()
   ...: print B
[[ 2. -1.  0.  0.  0.  0.]
 [-1.  2. -1.  0.  0.  0.]
 [ 0. -1.  2. -1.  0.  0.]
 [ 0.  0. -1.  2. -1.  0.]
 [ 0.  0.  0. -1.  2. -1.]
 [ 0.  0.  0.  0. -1.  2.]]
In [9]: np.set_printoptions(suppress=True, precision=3)
In [10]: print spla.cholesky(B)
[[ 1.414 -0.707  0.     0.     0.     0.    ]
 [ 0.     1.225 -0.816  0.     0.     0.    ]
 [ 0.     0.     1.155 -0.866  0.     0.    ]
 [ 0.     0.     0.     1.118 -0.894  0.    ]
 [ 0.     0.     0.     0.     1.095 -0.913]
 [ 0.     0.     0.     0.     0.     1.08 ]]
In [11]: print spla.cho_factor(B)[0]
[[ 1.414 -0.707  0.     0.     0.     0.    ]
 [-1.     1.225 -0.816  0.     0.     0.    ]
```

```
    [ 0.    -1.     1.155 -0.866  0.     0.   ]
    [ 0.     0.    -1.     1.118 -0.894  0.   ]
    [ 0.     0.     0.    -1.     1.095 -0.913]
    [ 0.     0.     0.     0.    -1.     1.08 ]]
In [12]: print spla.cholesky_banded(B_banded)
[[ 0.    -0.707 -0.816 -0.866 -0.894 -0.913]
 [ 1.414  1.225  1.155  1.118  1.095  1.08 ]]
```

The output of `cho_factor` is a tuple: the second element is the Boolean lower. The first element is ndarray representing a square matrix. If `lower` is set to `True`, the lower triangular sub-matrix of this ndarray is `L` in the Cholesky factorization of `A`. If `lower` is set to `False`, the upper triangular sub-matrix is `U` in the factorization of `A`. The remaining elements in the matrix are random, instead of zeros, since they are not used by `cho_solve`. In a similar way, we can call `cho_solve_banded` with the output of `cho_banded` to solve the appropriate system.

Both `cholesky` and `cho_factor` are wrappers to the same LAPACK function called `potrf`, with different output options. `cholesky_banded` calls `pbtrf`. The `cho_solve` function is a wrapper for `potrs`, and `cho_solve_banded` calls `pbtrs`.

We are then ready to solve the system, with either of the two options:

```
In [13]: spla.cho_solve((spla.cholesky(B), False), np.ones(6))
Out[13]: array([ 3.,  5.,  6.,  6.,  5.,  3.])
In [13]: spla.cho_solve(spla.cho_factor(B), np.ones(6))
Out[13]: array([ 3.,  5.,  6.,  6.,  5.,  3.])
```

For any other kind of generic square matrix A, the next best method to solve the basic system $A \cdot x = b$ is pivoted `LU` factorization. This is equivalent to finding a permutation matrix `P`, and triangular matrices `U` (upper) and `L` (lower) so that $P \cdot A = L \cdot U$. In such a case, a permutation of the rows in the system according to `P` gives the equivalent equation $(P \cdot A) \cdot x = P \cdot b$. Set `c = P · b` and `y = U · x`, and solve for `y` in the system $L \cdot y = c$ using forward substitution. Then, solve for `x` in the system $U \cdot x = y$ with back substitution.

The relevant functions to perform this operation are `lu`, `lu_factor` (for factorization), and `lu_solve` (for solution) in the module `scipy.linalg`. For sparse matrices we have `splu`, and `spilu`, in the module `scipy.sparse.linalg`.

Let us start experimenting with factorizations first. We use a large circulant matrix (non-symmetric) for this example:

```
In [14]: D = spla.circulant(np.arange(4096))
In [15]: %timeit spla.lu(D)
1 loops, best of 3: 7.04 s per loop
In [16]: %timeit spla.lu_factor(D)
1 loops, best of 3: 5.48 s per loop
```

 The lu_factor function is a wrapper to all *getrf routines from LAPACK. The lu_solve function is a wrapper for getrs.

The function lu has an extra Boolean option: permute_l (set to False by default). If set to True, the function outputs only two matrices $PL = P \cdot L$ (the properly permuted lower triangular matrix), and U. Otherwise, the output is the triple P, L, U, in that order.

```
In [17]: P, L, U = spla.lu(D)
In [17]: PL, U = spla.lu(D, permute_l=True)
```

The outputs of the function lu_factor are resource-efficient. We obtain a matrix LU, with upper triangle U and lower triangle L. We also obtain a one-dimensional ndarray class of integer dtype, piv, indicating the pivot indices representing the permutation matrix P.

```
In [18]: LU, piv = spla.lu_factor(D)
```

The solver lu_solve takes the two outputs from lu_factor, a right-hand side matrix b, and the optional indicator trans to the kind of basic system to solve:

```
In [19]: spla.lu_solve(spla.lu_factor(D), np.ones(4096))
Out[19]: array([ 0.,   0.,   0., ...,   0.,   0.,   0.])
```

 At this point, we must comment on the general function solve in the module scipy.linalg. It is a wrapper to both LAPACK functions POSV and GESV. It allows us to input matrix A and right-hand side matrix b, and indicate whether A is symmetric and positive definite. In any case, the routine internally decides which of the two factorizations to use (Cholesky or pivoted LU), and computes a solution accordingly.

For large sparse matrices, provided they are stored in the CSC format, the pivoted LU decomposition is more efficiently performed with either functions `splu` or `spilu` from the module `scipy.sparse.linalg`. Both functions use the `SuperLU` library directly. Their output is not a set of matrices, but a Python object called `scipy.sparse.linalg.dsolve._superlu.SciPyLUType`. This object has four attributes and one instance method:

- `shape`: 2-tuple containing the shape of matrix `A`

- `nnz`: The number of non-zero entries in matrix `A`

- `perm_c`, `perm_r`: The permutations applied to the columns and rows (respectively) to the matrix `A` to obtain the computed LU decomposition

- `solve`: instance method that converts the object into a function `object.solve(b,trans)` accepting `ndarray` `b`, and the optional description string `trans`.

The big idea is that, dealing with large amounts of data, the actual matrices in the LU decomposition are not as important as the main application behind the factorization: the solution of the system. All the relevant information to perform this operation is optimally stored in the object's method `solve`.

The main difference between `splu` and `spilu` is that the latter computes an incomplete decomposition. With it, we can obtain really good approximations to the inverse of matrix `A`, and use matrix multiplication to compute the solution of large systems in a fraction of the time that it would take to calculate the actual solution.

The usage of these two functions is rather complex. The purpose is to compute a factorization of the form $Pr*Dr*A*Dc*Pc = L*U$ with diagonal matrices `Dr` and `Dc` and permutation matrices `Pr` and `Pc`. The idea is to equilibrate matrix `A` manually so that the product $B = Dr*A*Dc$ is better conditioned than `A`. In case of the possibility of solving this problem in a parallel architecture, we are allowed to help by rearranging the rows and columns optimally. The permutation matrices `Pr` and `Pc` are then manually input to pre-order the rows and columns of `B`. All of these options can be fed to either `splu` or `spilu`.

The algorithm exploits the idea of relaxing supernodes to reduce inefficient indirect addressing and symbolic time (besides permitting the use of higher-level BLAS operations). We are given the option to determine the degree of these objects, to tailor the algorithm to the matrix at hand.

For a complete explanation of the algorithms and all the different options, the best reference is SuperLU User Guide, which can be found online at `crd-legacy.lbl.gov/~xiaoye/SuperLU/superlu_ug.pdf`.

Let us illustrate this with a simple example, where the permutation of rows or columns is not needed. In a large lower triangular Pascal matrix, turn into zero all the even-valued entries and into ones all the odd-valued entries. Use this as matrix A. For the right-hand side, use a vector of ones:

```
In [20]: A_csc = spsp.csc_matrix(A, dtype=np.float64)
In [21]: invA = spspla.splu(A_csc)
In [22]: %time invA.solve(np.ones(1024))
CPU times: user: 4.32 ms, sys: 105 µs, total: 4.42 ms
Wall time: 4.44 ms
Out[22]: array([ 1., -0.,  0., ..., -0.,  0.,  0.])
In [23]: invA = spspla.spilu(A_csc)
In [24]: %time invA.solve(np.ones(1024))
CPU times: user 656 µs, sys: 22 µs, total: 678 µs
Wall time: 678 µs
Out[24]: array([ 1.,  0.,  0., ...,  0.,  0.,  0.])
```

 Compare the time of execution of the procedures on sparse matrices, with the initial `solve_triangular` procedure on the corresponding matrix A at the beginning of the section. Which process is faster?

However, in general, if a basic system must be solved and matrix A is large and sparse, we prefer to use iterative methods with fast convergence to the actual solutions. When they converge, they are consistently less sensitive to rounding-off errors and thus more suitable when the number of computations is extremely high.

In the module `scipy.sparse.linalg`, we have eight different iterative methods, all of which accept the following as parameters:

- Matrix A in any format (matrix, ndarray, sparse matrix, or even a linear operator!), and right-hand side vector/matrix b as `ndarray`.

- Initial guess x0, as `ndarray`.

- Tolerance to 1, a floating point number. If the difference of successive iterations is less than this value, the code stops and the last computed values are output as the solution.

- Maximum number of iterations allowed, maxiter, an integer.

- A Preconditioner sparse matrix M that should approximate the inverse of A.

- A `callback` function of the current solution vector xk, called after each iteration.

Constructor	Description
bicg	Biconjugate Gradient Iteration
bicgstab	Biconjugate Gradient Stabilized Iteration
cg	Conjugate Gradient Iteration
cgs	Conjugate Gradient Squared Iteration
gmres	Generalized Minimal Residual Iteration
lgmres	LGMRES Iteration
minres	Minimum Residual Iteration
qmr	Quasi-minimal Residual Iteration

Choosing the right iterative method, a good initial guess, and especially a successful Preconditioner is an art in itself. It involves learning about topics such as operators in Functional Analysis, or Krylov subspace methods, which are far beyond the scope of this book. At this point, we are content with showing a few simple examples for the sake of comparison:

```
In [25]: spspla.cg(A_csc, np.ones(1024), x0=np.zeros(1024))
Out[25]: (array([ nan,  nan,  nan, ...,  nan,  nan,  nan]), 1)
In [26]: %time spspla.gmres(A_csc, np.ones(1024), x0=np.zeros(1024))
CPU times: user 4.26 ms, sys: 712 µs, total: 4.97 ms
Wall time: 4.45 ms
Out[26]: (array([ 1.,  0.,  0., ..., -0., -0.,  0.]), 0)
In [27]: Nsteps = 1
   ....: def callbackF(xk):
   ....:     global Nsteps
   ....:     print'{0:4d}  {1:3.6f}  {2:3.6f}'.format(Nsteps, \
   ....:     xk[0],xk[1])
   ....:     Nsteps += 1
   ....:
In [28]: print '{0:4s}  {1:9s}  {1:9s}'.format('Iter', \
   ....: 'X[0]','X[1]'); \
   ....: spspla.bicg(A_csc, np.ones(1024), x0=np.zeros(1024),
   ....:             callback=callbackF)
   ....:
```

```
Iter  X[0]        X[1]
   1  0.017342  0.017342
   2  0.094680  0.090065
   3  0.258063  0.217858
   4  0.482973  0.328061
   5  0.705223  0.337023
   6  0.867614  0.242590
   7  0.955244  0.121250
   8  0.989338  0.040278
   9  0.998409  0.008022
  10  0.999888  0.000727
  11  1.000000  -0.000000
  12  1.000000  -0.000000
  13  1.000000  -0.000000
  14  1.000000  -0.000000
  15  1.000000  -0.000000
  16  1.000000  0.000000
  17  1.000000  0.000000
Out[28]: (array([ 1.,   0.,   0., ...,   0.,   0.,  -0.]), 0)
```

Least squares

Given a generic matrix A (not necessarily square) and a right-hand side vector/matrix b, we look for a vector/matrix x such that the Frobenius norm of the expression $A \cdot x - b$ is minimized.

The main three methods to solve this problem numerically are contemplated in scipy:

- Normal equations
- QR factorization
- Singular value decomposition

Normal equations

Normal equations reduce the least square problem to solving a basic system of linear equations, with a symmetric (not-necessarily positive-definite) matrix. It is very fast but can be inaccurate due to presence of roundoff errors. Basically, it amounts to solving the system $(A^H \cdot A) \cdot x = A^H \cdot b$. This is equivalent to solving $x = (A^H \cdot A)^{-1} \cdot A^H \cdot b = pinv(A) \cdot b$.

Let us show by example:

```
In [29]: E = D[:512,:256]; b = np.ones(512)
In [30]: sol1 = np.dot(spla.pinv2(E), b)
In [31]: sol2 = spla.solve(np.dot(F.T, F), np.dot(F.T, b))
```

QR factorization

The QR factorization turns any matrix into the product $A = Q \cdot R$ of an orthogonal/ unitary matrix Q with a square upper triangular matrix R. This allows us to solve the system without the need to invert any matrix (since $Q^H = Q^{-1}$), and thus, $A \cdot x = b$ turns into $R \cdot x = Q^H \cdot b$, which is easily solvable by back substitution. Note that the two methods below are equivalent, since the mode economic reports the sub-matrices of maximum rank:

```
In [32]: Q, R = spla.qr(E); \
    ....: RR = R[:256, :256]; BB = np.dot(Q.T, b)[:256]; \
    ....: sol3 = spla.solve_triangular(RR, BB)
In [32]: Q, R = spla.qr(E, mode='economic'); \
    ....: sol3 = spla.solve_triangular(R, np.dot(Q.T, b))
```

Singular value decomposition

Both methods of normal equations and QR factorization work fast and are reliable only when the rank of A is full. If this is not the case, we must use singular value decomposition $A = U \cdot D \cdot V^H$ with unitary matrices U and V and a diagonal matrix D, where all the entries in the diagonal are positive values. This allows for a fast solution $x = V \cdot D^{-1} \cdot U^H \cdot b$.

Note that the two methods discussed below are equivalent, since the option full_matrices set to False reports the sub-matrices of the minimum possible size:

```
In [33]: U, s, Vh = spla.svd(E); \
    ....: Uh = U.T; \
    ....: Si = spla.diagsvd(1./s, 256, 256); \
    ....: V = Vh.T; \
    ....: sol4 = np.dot(V, Si).dot(np.dot(Uh, b)[:256])
In [33]: U, s, Vh = spla.svd(E, full_matrices=False); \
    ....: Uh = U.T; \
    ....: Si = spla.diagsvd(1./s, 256, 256); \
    ....: V = Vh.T; \
    ....: sol4 = np.dot(V, Si).dot(np.dot(Uh, b))
```

The module `scipy.linalg` has one function that actually performs least squares with the SVD method: `lstsq`. There is no need to manually transpose, invert, and multiply all the required matrices. It is a wrapper to the LAPACK function GELSS. It outputs the desired solution, together with the residues of computation, the effective rank, and the singular values of the input matrix A.

```
In [34]: sol5, residue, rank, s = spla.lstsq(E, b)
```

Note how all the computations that we have carried out offer solutions that are very close to each other (if not equal!):

```
In [35]: map(lambda x: np.allclose(sol5,x), [sol1, sol2, sol3, sol4])
Out[35]: [True, True, True, True]
```

Regularized least squares

The module `scipy.sparse.linalg` has two iterative methods for least squares in the context of large sparse matrices, `lsqr` and `lsmr`, which allow for a more generalized version with a damping factor `d` for regularization. We seek to minimize the functional `norm(A * x - b, 'f')**2 + d**2 * norm(x, 'f')**2`. The usage and parameters are very similar to the iterative functions we studied before.

Other matrix equation solvers

The rest of the matrix equation solvers are summarized in the following table. None of these routines enjoy any parameters to play around with performance or memory management, or check for the integrity of data:

Constructor	Description
`solve_sylvester(A, B, Q)`	Sylvester equation
`solve_continuous_are(A, B, Q, R)`	continuous algebraic Riccati equation
`solve_discrete_are(A, B, Q, R)`	discrete algebraic Riccati equation
`solve_lyapunov(A, Q)`	continuous Lyapunov equation
`solve_discrete_lyapunov(A, Q)`	discrete Lyapunov equation

Matrix factorizations based on eigenvalues

In this category, we have two kinds of factorizations on square matrices: Spectral and Schur decompositions (although, technically, a spectral decomposition is a special case of Schur decomposition). The objective of both is initially to present the eigenvalues of one or several matrices simultaneously, although they have quite different applications.

Spectral decomposition

We consider the following four cases:

- Given a square matrix A, we seek all vectors v (right eigenvectors) that satisfy $A \cdot v = m \cdot v$ for some real or complex value m (the corresponding eigenvalues). If all eigenvectors are different, we collect them as the columns of matrix V (that happens to be invertible). Their corresponding eigenvalues are stored in the same order as the diagonal entries of a diagonal matrix D. We can then realize A as the product $A = V \cdot D \cdot V^{-1}$. We refer to this decomposition as an ordinary eigenvalue problem.

- Given a square matrix A, we seek all vectors v (left eigenvectors) that satisfy $v \cdot A = m \cdot v$ for the eigenvalues m. As before, if all eigenvectors are different, they are collected in matrix V; their corresponding eigenvalues are collected in the diagonal matrix D. The matrix A can then be decomposed as the product $A = V \cdot D \cdot V^{-1}$. We also refer to this factorization as an ordinary eigenvalue problem. The eigenvalues are the same as in the previous case.

- Given square matrices A and B with the same size, we seek all vectors v (generalized right eigenvectors) that satisfy $m \cdot A \cdot v = n \cdot B \cdot v$ for some real or complex values m and n. The ratios $r = n/m$, when they are computable, are called generalized eigenvalues. The eigenvectors are collected as columns of matrix V, and their corresponding generalized eigenvalues r collected in a diagonal matrix D. We can then realize the relation between A and B by the identity $A = B \cdot V \cdot D \cdot V^{-1}$. We refer to this identity as a generalized eigenvalue problem.

- For the same case as before, if we seek vectors v (generalized left eigenvectors) and values m and n that satisfy $m \cdot v \cdot A = n \cdot v \cdot B$, we have another similar decomposition. We again refer to this factorization as a generalized eigenvalue problem.

The following functions in the modules `scipy.linalg` and `scipy.sparse.linalg` help us to compute eigenvalues and eigenvectors:

Constructor	Description
`scipy.linear.eig(A[, B])`	Ordinary/generalized eigenvalue problem
`scipy.linalg.eigvals(A[, B])`	Eigenvalues for ordinary/generalized eigenvalue problem
`scipy.linalg.eigh(A[, B])`	Ordinary/generalized eigenvalue problem. Hermitian/symmetric matrix
`scipy.linalg.eigvalsh(A[, B])`	Eigenvalues for ordinary/generalized eigenvalue problem; Hermitian/symmetric matrix
`scipy.linalg.eig_banded(AB)`	Ordinary eigenvalue problem; Hermitian/symmetric band matrix
`scipy.linalg.eigvals_banded(AB)`	Eigenvalues for ordinary eigenvalue problem; Hermitian/symmetric band matrix
`scipy.sparse.linalg.eigs(A, k)`	Find k eigenvalues and eigenvectors
`scipy.sparse.linalg.eigsh(A, k)`	Find k eigenvalues and eigenvectors; Real symmetric matrix
`scipy.sparse.linalg.lobpcg(A, X)`	Ordinary/generalized eigenvalue problem with optional preconditioning A symmetric

For any kind of eigenvalue problem where the matrices are not symmetric or banded, we use the function `eig`, which is a wrapper for the LAPACK routines GEEV and GGEV (the latter for generalized eigenvalue problems). The function `eigvals` is syntactic sugar for a case of `eig` that only outputs the eigenvalues, but not the eigenvectors. To report whether we require left of right eigenvectors, we use the optional Boolean parameters `left` and `right`. By default, `left` is set to `False` and `right` to `True`, hence offering right eigenvectors.

For eigenvalue problems with non-banded real symmetric or Hermitian matrices, we use the function `eigh`, which is a wrapper for the LAPACK routines of the form *EVR, *GVD, and *GV. We are given the choice to output as many eigenvalues as we want, with the optional parameter `eigvals`. This is a tuple of integers that indicate the indices of the lowest and the highest eigenvalues required. If omitted, all eigenvalues are returned. In such a case, it is possible to perform the computation with a much faster algorithm based on divide and conquer techniques. We may indicate this choice with the optional Boolean parameter `turbo` (by default set to `False`).

If we wish to report only eigenvalues, we can set the optional parameter eigvals_only to True, or use the corresponding syntactic sugar eighvals.

The last case that we contemplate in the scipy.linalg module is that of the eigenvalue problem of a banded real symmetric or Hermitian matrix. We use the function eig_banded, making sure that the input matrices are in the AB format. This function is a wrapper for the LAPACK routines *EVX.

For extremely large matrices, the computation of eigenvalues is often computationally impossible. If these large matrices are sparse, it is possible to calculate a few eigenvalues with two iterative algorithms, namely the **Implicitly Restarted Arnoldi** and the **Implicitly Restarted Lanczos** methods (the latter for symmetric or Hermitian matrices). The module scipy.sparse.linalg has two functions, eigs and eigsh, which are wrappers to the ARPACK routines *EUPD that perform them. We also have the function lobpcg that performs another iterative algorithm, the **Locally Optimal Block Preconditioned Conjugate Gradient** method. This function accepts a Preconditioner, and thus has the potential to converge more rapidly to the desired eigenvalues.

We will illustrate the usage of all these functions with an interesting matrix: Andrews. It was created in 2003 precisely to benchmark memory-efficient algorithms for eigenvalue problems. It is a real symmetric sparse matrix with size 60,000 × 60,000 and 760,154 non-zero entries. It can be downloaded from the Sparse Matrix Collection at www.cise.ufl.edu/research/sparse/matrices/Andrews/Andrews.html.

For this example, we downloaded the matrix in the Matrix Market format Andrews.mtx. Note that the matrix is symmetric, and the file only provides data on or below the main diagonal. After collecting all this information, we ensure that we populate the upper triangle too:

```
In [1]: import numpy as np, scipy.sparse as spsp, \
   ...: scipy.sparse.linalg as spspla
In [2]: np.set_printoptions(suppress=True, precision=6)
In [3]: rows, cols, data = np.loadtxt("Andrews.mtx", skiprows=14,
   ...:                                   unpack=True); \
   ...: rows-=1; \
   ...: cols-=1
In [4]: A = spsp.csc_matrix((data, (rows, cols)), \
   ...:                     shape=(60000,60000)); \
   ...: A = A + spsp.tril(A, k=1).transpose()
```

We compute first the top largest five eigenvalues in absolute value. We call the function `eigsh`, with the option `which='LM'`.

```
In [5]: %time eigvals, v = spspla.eigsh(A, 5, which='LM')
CPU times: user 3.59 s, sys: 104 ms, total: 3.69 s
Wall time: 3.13 s
In [6]: print eigvals
[ 69.202683   69.645958   70.801108   70.815224   70.830983]
```

We may compute the smallest eigenvalues in terms of the absolute value too, by switching to the option `which='SM'`:

```
In [7]: %time eigvals, v = spspla.eigsh(A, 5, which='SM')
CPU times: user 19.3 s, sys: 532 ms, total: 19.8 s
Wall time: 16.7 s
In [8]: print eigvals
[ 10.565523   10.663114   10.725135   10.752737   10.774503]
```

The routines in ARPACK are not very efficient at finding small eigenvalues. It is usually preferred to apply the shift-invert mode in this case for better performance. For information about this procedure, read the description in www.caam.rice.edu/software/ARPACK/UG/node33.html, or the article by R. B. Lehoucq, D. C. Sorensen, and C. Yang, ARPACK USER GUIDE: *Solution of Large Scale Eigenvalue Problems by Implicitly Restarted Arnoldi Methods*. SIAM, Philadelphia, PA, 1998.

The function `eigsh` allows us to perform shift-invert mode by indicating a value close to the required eigenvalues. If we have a good guess, as offered by the previous step, we may apply this procedure with the option `sigma`, and a strategy with the option `mode`. In this case, we also need to provide a linear operator instead of a matrix. The time of execution is much slower, but the results are much more precise in general (although the given example would not suggest so!).

```
In [9]: A = spspla.aslinearoperator(A)
In [10]: %time spspla.eigsh(A, 5, sigma=10.0, mode='cayley')
CPU times: user 2min 5s, sys: 916 ms, total: 2min 6s
Wall time: 2min 6s
In [11]: print eigvals
[ 10.565523   10.663114   10.725135   10.752737   10.774503]
```

Schur decomposition

There are four cases:

- **Complex Schur decomposition** for a square matrix A with complex coefficients. We can realize A as the product $A = U \cdot T \cdot U^H$ of a unitary matrix U with an upper triangular matrix T, and the Hermitian transpose of U. We call T the complex Schur form of A. The entries in the diagonal of T are the eigenvalues of A.

- **Real Schur decomposition** for a square matrix A with real coefficients. If all the eigenvalues of the matrix are real valued, then we may realize the matrix as the product $A = V \cdot S \cdot V^T$ of an orthonormal matrix V with a block-upper triangular matrix S, and the transpose of V. The blocks in S are either of size 1×1 or 2×2. If the block is 1×1, the value is one of the real eigenvalues of A. Any 2×2 blocks represents a pair of complex conjugate eigenvalues of A. We call S the real Schur form of A.

- **Complex generalized Schur decomposition** of two square matrices A and B. We can simultaneously factorize them to the form $A = Q \cdot S \cdot Z^H$ and $B = Q \cdot T \cdot Z^H$ with the same unitary matrices Q and Z. The matrices S and T are both upper triangular, and the ratios of their diagonal elements are precisely the generalized eigenvalues of A and B.

- **Real generalized Schur decomposition** of two real-valued square matrices A and B. Simultaneous factorization of both can be achieved in the form $A = Q \cdot S \cdot Z^T$ and $B = Q \cdot T \cdot Z^T$ for the same orthogonal matrices Q and Z. The matrices S and T are block-upper triangular, with blocks of size 1×1 and 2×2. With the aid of these blocks, we can find the generalized eigenvalues of A and B.

There are four functions in the module `scipy.linalg` that provide us with tools to compute any of these decompositions:

Constructor	Description
`scipy.linalg.schur(A)`	Schur decomposition of a matrix
`scipy.linalg.rsf2csf(T, Z)`	Convert from real Schur form to complex Schur form
`scipy.linalg.qz(A, B)`	Generalized Schur decomposition of two matrices
`scipy.linalg.hessenberg(A)`	Hessenberg form of a matrix

The function `hessenberg` gives us the first step in the computation of any Schur decomposition. This is a factorization of any square matrix A in the form $A = Q \cdot U \cdot Q^H$, where Q is unitary and U is an upper Hessenberg matrix (all entries are zero below the sub-diagonal). The algorithm is based on the combination of the LAPACK routines GEHRD, GEBAL (to compute U), and the BLAS routines GER, GEMM (to compute Q).

The functions `schur` and `qz` are wrappers to the LAPACK routines GEES and GGES, to compute the normal and generalized Schur decompositions (respectively) of square matrices. We choose whether to report complex or real decompositions on the basis of the optional parameter output (which we set to `'real'` or `'complex'`). We also have the possibility of sorting the eigenvalues in the matrix representation. We do so with the optional parameter `sort`, with the following possibilities:

- `None`: If we do not require any sorting. This is the default.
- `'lhp'`: In the left-hand plane.
- `'rhp'`: In the right-hand plane
- `'iuc'`: Inside the unit circle
- `'ouc'`: Outside the unit circle
- `func`: Any callable function called `func` can be used to provide the users with their own sorting

Summary

In this chapter, we have explored the basic principles of numerical linear algebra — the core of all procedures in scientific computing. The emphasis was first placed on the storage and the basic manipulation of matrices and linear operators. We explored in detail all different factorizations, focusing on their usage to find a solution to matrix equations or eigenvalue problems. All through the chapter, we made it a point to link the functions from the modules `scipy.linalg` and `scipy.sparse` to their corresponding routines in the libraries BLAS, LAPACK, ARPACK and SuperLU. For our experiments, we chose interesting matrices from real-life problems that we gathered from the extensive Sparse Matrix Collection hosted by the University of Florida.

In the next chapter, we will address the problems of interpolation and least squares approximation.

2

Interpolation and Approximation

Approximation theory states how to find the best approximation to a given function by another function from some predetermined class and how good this approximation is. In this chapter, we are going to explore this field through two settings: **interpolation** and **least squares approximation**.

Motivation

Consider a meteorological experiment that measures the temperature of a set of buoys located on a rectangular grid at sea. We can emulate such an experiment by indicating the longitude and latitude of the buoys on a grid of 16×16 locations, and random temperatures on them between say 36°F and 46°F:

```
In [1]: import numpy as np, matplotlib.pyplot as plt, \
   ...: matplotlib.cm as cm; \
   ...: from mpl_toolkits.basemap import Basemap
In [2]: map1 = Basemap(projection='ortho', lat_0=20, lon_0=-60, \
   ...:                 resolution='l', area_thresh=1000.0); \
   ...: map2 = Basemap(projection='merc', lat_0=20, lon_0=-60, \
   ...:                 resolution='l', area_thresh=1000.0, \
   ...:                 llcrnrlat=0,  urcrnrlat=45, \
   ...:                 llcrnrlon=-75, urcrnrlon=-15)
In [3]: longitudes = np.linspace(-60, -30, 16); \
   ...: latitudes = np.linspace(15, 30, 16); \
   ...: lons, lats = np.meshgrid(longitudes, latitudes); \
   ...: temperatures = 10. * np.random.randn(16, 16) + 36.
```

```
In [4]: x1, y1 = map1(lons, lats); \
   ...: x2, y2 = map2(lons, lats)
In [5]: plt.rc('text', usetex=True); \
   ...: plt.figure()
In [6]: plt.subplot(121, aspect='equal'); \
   ...: map1.drawmeridians(np.arange(0, 360, 30)); \
   ...: map1.drawparallels(np.arange(-90, 90, 15)); \
   ...: map1.drawcoastlines(); \
   ...: map1.fillcontinents(color='coral'); \
   ...: map1.scatter(x1, y1, 15, temperatures, cmap=cm.gray)
In [7]: plt.subplot(122); \
   ...: map2.drawmeridians(np.arange(0, 360, 30)); \
   ...: map2.drawparallels(np.arange(-90, 90, 15)); \
   ...: map2.drawcoastlines(); \
   ...: map2.fillcontinents(color='coral'); \
   ...: C = map2.scatter(x2, y2, 15, temperatures, cmap=cm.gray); \
   ...: Cb = map2.colorbar(C, "bottom", size="5%", pad="2%"); \
   ...: Cb.set_label(r'$\mbox{}^{\circ} F$'); \
   ...: plt.show()
```

We obtain the following diagram:

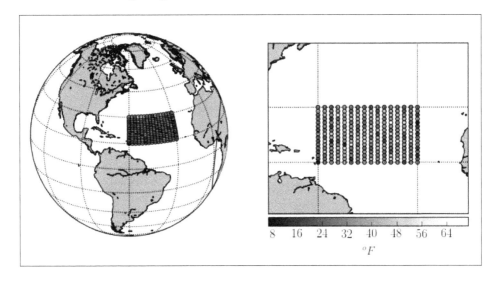

It is possible to guess the temperature between these buoys (not exactly, but at least to some extent), since the temperature is a smooth function on the surface of the Earth. Let us assume that we require an approximation by means of piecewise 2D polynomials of degree three, with maximum smoothness where the pieces intersect each other. One obvious challenge is, of course, that the buoys are not located on a plane, but on the surface of a really big sphere. This is not an issue for SciPy.

```
In [8]: from scipy.interpolate import RectSphereBivariateSpline \
   ...: as RSBS
In [9]: soln = RSBS(np.radians(latitudes), \
   ...:             np.pi + np.radians(longitudes), \
   ...:             temperatures)
In [10]: long_t = np.linspace(-60, -30, 180); \
   ....: lat_t = np.linspace(15, 30, 180); \
   ....: temperatures = soln(np.radians(lat_t), \
   ....:                    np.pi + np.radians(long_t))
In [11]: long_t, lat_t = np.meshgrid(long_t, lat_t); \
   ....: lo1, la1 = map1(long_t, lat_t); \
   ....: lo2, la2 = map2(long_t, lat_t)
In [12]: plt.figure()
Out[12]: <matplotlib.figure.Figure at 0x10ec28250>
In [13]: plt.subplot(121, aspect='equal'); \
   ....: map1.drawmeridians(np.arange(0, 360, 30)); \
   ....: map1.drawparallels(np.arange(-90, 90, 15)); \
   ....: map1.drawcoastlines(); \
   ....: map1.fillcontinents(color='coral'); \
   ....: map1.contourf(lo1, la1, temperatures, cmap=cm.gray)
Out[13]: <matplotlib.contour.QuadContourSet instance at 0x10f63d7e8>
In [14]: plt.subplot(122); \
   ....: map2.drawmeridians(np.arange(0, 360, 30)); \
   ....: map2.drawparallels(np.arange(-90, 90, 15)); \
   ....: map2.drawcoastlines(); \
   ....: map2.fillcontinents(color='coral'); \
   ....: C = map2.contourf(lo2, la2, temperatures, cmap=cm.gray); \
   ....: Cb = map2.colorbar(C, "bottom", size="5%", pad="2%"); \
```

```
....: Cb.set_label(r'$\mbox{}^{\circ} F$'); \
....: plt.show()
```

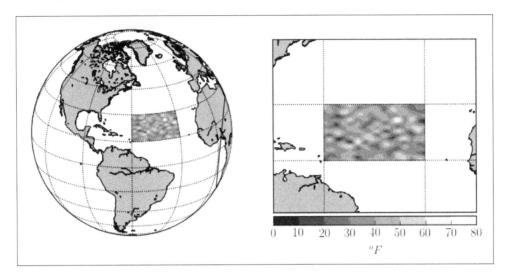

We have solved this problem simply by seeking a representation of the temperature function as a piecewise polynomial surface that agrees with the temperature values over the locations of the buoys. This is technically called an interpolation with bivariate splines over a rectangular grid of nodes on a sphere.

In other situations, the precise value at these locations is not very important, provided that the resulting function is more closely related to the actual temperature. In such a case, rather than performing interpolation, we want to compute an approximation with the elements of the same functional class.

Let us define both the settings precisely:

The interpolation problem requires three ingredients:

- A target function f(x) on a finite domain (which we denote by x for convenience).

- A finite set of points in the domain: The nodes of interpolation, which we denote by xi. We will also need the evaluation of the target function (and possibly some of its derivatives) at these nodes. We denote these by yi throughout this chapter.

- A family of interpolants: Functions with the same input/output structure as the target function.

The goal of the interpolation problem is the approximation of the target function by a member of the interpolants by matching the values of the target function at the nodes.

We explore interpolation in the following settings:

- Nearest-neighbors interpolation (in any dimension)
- Interpolation by piecewise linear functions (in any dimension)
- Univariate interpolation by polynomials (both Lagrange and Hermite interpolation)
- Univariate interpolation by piecewise polynomials
- Univariate and bivariate interpolation by splines
- Radial basis multivariate interpolation

We assume familiarity with the theory and application of splines. There are many good sources to get started, but we recommend those with a more practical flavor:

- Carl de Boor, *A Practical Guide to Splines*. Springer, 1978.
- Paul Dierckx, *Curve and Surface Fitting with Splines*. Oxford University Press, 1993.

All the interpolations are carried out through the module `scipy.interpolate`. In particular, those related to splines are a set of wrappers to some routines in Paul Dierckx's `FITPACK` libraries.

To define the approximation problem, we need the following four ingredients:

- A target function `f(x)` on a finite domain `x` that takes as input a column vector of dimension `n` and outputs a column vector of dimension `m`
- A family of approximants, `{g[a](x)}`: Functions with the same input/output structure as `f(x)`, that depend on parameter `a` coded as a column vector of dimension `r`
- A `norm` is a functional that measures the distance between any two given functions of `x`, `||f(x) - g(x)||`

The goal of the approximation problem is to find the member of the approximants that minimizes the expression `||f(x) - g[a](x)||` with respect to parameter `a`. This is equivalent to finding `a` (local or global) minimum for `error` function `F(a) = ||f(x) - g[a](x)||` over `a`.

We say that the approximation is **linear** if the family of approximants is a linear combination of elements of a basis and the parameter a acts as the coefficient; otherwise, we refer to the approximation as **nonlinear**.

In this chapter, we will address the approximation of functions in the following settings:

- Generic linear least squares approximation (by solving systems of linear equations)
- Least squares approximation/smoothing with univariate and bivariate splines
- Least squares approximation/smoothing with splines over rectangular grids on spheres
- Generic nonlinear least squares approximation (with the Levenberg-Marquardt iterative algorithm)

Least squares approximation in the context of functions is performed through several modules:

- For a generic linear least squares approximation, the problems can always be reduced to solutions of systems of linear equations. In this case, the `scipy.linalg` and `scipy.sparse.linalg` modules that we studied in the previous chapter hold all the algorithms that we need. As explained earlier, the required functions are wrappers of several routines in the Fortran libraries BLAS and LAPACK, and the C library SuperLU.
- For the special case of linear least squares approximation through splines, the `scipy.interpolate` module carries out many functions (for all the different cases), which are in turn wrappers of routines in Paul Diercks's Fortran library FITPACK.
- For a nonlinear least squares approximation, we use functions from the `scipy.optimize` module. These functions are wrappers to the LMDIF and LMDER routines in the Fortran library MINPACK.

For more information on these Fortran libraries, a good reference can be obtained from their pages in the Netlib repository:

- FITPACK: `http://netlib.org/dierckx/`
- MINPACK: `http://netlib.org/minpack/`
- FFTPACK: `http://netlib.org/fftpack/`

One of the best references for SuperLU can be found from its creators at `http://crd-legacy.lbl.gov/~xiaoye/SuperLU/`.

Interpolation

We have three different implementation methodologies to deal with interpolation problems:

- A **procedural** mode that computes a set of data points (in the form of `ndarray` with the required dimension) representing the actual solution.

- In a few special cases, a **functional** mode that provides us with `numpy` functions representing the solutions.

- An object-oriented mode that creates classes for interpolation problems. Different classes have different methods, depending on the operations that the particular kinds of interpolants enjoy. The advantage of this mode is that, through these methods, we can request more information from the solutions: not only evaluation or representation, but also relevant operations like searching for roots, computing derivatives and antiderivatives, error checking, and calculating coefficients and knots.

The choice of mode to represent our interpolants is up to us, depending mostly on how much accuracy we require, and the information/operations that we need afterwards.

Implementation details

There is not much more to add to the implementation details on the procedural mode. For each interpolation problem, we choose a routine to which we feed nodes `xi`, the values `yi` of the target function (and possibly its derivatives) on those nodes, and the domain `x` where the interpolant is to be evaluated. In some cases, if the interpolant requires more structure, we feed extra information.

Functional implementations are even simpler: when they are available, they only require the values of nodes `xi` and evaluation at those nodes, `yi`.

There are several generic object-oriented classes for interpolation. We seldom tamper with them and resort to using routines that create and manipulate internally more appropriate subclasses instead. Let us go over these objects in brief:

- For generic univariate interpolation, we have the `_Interpolator1D` class. It may be initialized with the set of nodes `xi`, together with the value of the target function on those nodes, `yi`. If necessary, we may force the data type of `yi` as well, with the `._set_dtype` class method. In case we need to deal with the derivatives of an interpolator, we have the `_Interpolator1DWithDerivatives` subclass, with the extra class method `.derivatives` to compute the evaluations of the differentiations.

- For univariate interpolation with splines of a degree less than or equal to 5, we have the `InterpolatedUnivariateSpline` class, which is in turn a subclass of the `UnivariateSpline` class. They are both very rich classes, with plenty of methods not only to evaluate a spline or any of its derivatives, but also to compute the spline representations of its derivatives and antiderivatives. We have methods to compute a definite integral between two points, as well. There are also methods that return the position of the knots, the spline coefficients, residuals, or even the roots. We initialize objects in the `UnivariateSpline` class at least with the nodes `xi` and the values to fit on those nodes, `yi`. We may optionally initialize the object with the degrees of the spline.

- For bivariate interpolation with unstructured nodes (nodes not necessarily on a rectangular grid), one option is the `interp2d` class, which implements interpolation in two dimensions with bivariate splines of orders 1, 3, or 5. This class is initialized with the nodes and their evaluation.

- For bivariate spline interpolation with nodes on a rectangular grid, we have the `RectBivariateSpline` class (when used with the `s = 0` parameter), which is a subclass of the `BivariateSpline` class. In turn, `BivariateSpline` is a subclass of the base class `_BivariateSplineBase`. As its univariate counterpart, this is a very rich class with many methods for evaluation, extraction of nodes and coefficients, and computation of volume integrals or residuals.

- For multivariate interpolation, there is the `NDInterpolatorBase` class, with three subclasses: `NearestNDInterpolator` (for nearest-neighbors interpolation), `LinearNDInterpolator` (for piecewise linear interpolation), and `CloughTocher2DInterpolator` (that implements a piecewise cubic, C1 smooth, curvature-minimizing interpolant in two dimensions).

- For interpolation on a set of nodes on a rectangular grid over the surface of a sphere, there is the `RectSphereBivariateSpline` subclass of the `SphereBivariateSpline` class. We initialize it with the angles (`theta` and `phi`) representing the location of the nodes on the sphere, together with the corresponding evaluations.

- For multivariate interpolation with radial functions, we have the `Rbf` class. It is rather dry, since it only allows an evaluation method. It is initialized with nodes and evaluations.

Univariate interpolation

The following table summarizes the different univariate interpolation modes coded in SciPy, together with the processes that we may use to resolve them:

Interpolation mode	Object-oriented implementation	Procedural implementation
Nearest-neighbors	`interp1d(,kind='nearest')`	
Lagrange polynom.	`BarycentricInterpolator`	`barycentric_interpolate`
Hermite polynom.	`KroghInterpolator`	`krogh_interpolate`
Piecewise polynom.	`PiecewisePolynomial`	`piecewise_polynomial_interpolate`
Piecewise linear	`interp1d(,kind='linear')`	
Generic spline interpolation	`InterpolatedUnivariateSpline`	`splrep`
Zero-order spline	`interp1d(,kind='zero')`	
Linear spline	`interp1d(,kind='slinear')`	
Quadratic spline	`interp1d(,kind='quadratic')`	
Cubic spline	`interp1d(,kind='cubic')`	
PCHIP	`PchipInterpolator`	`pchip_interpolate`

Nearest-neighbors interpolation

In the context of one-dimensional functions, nearest-neighbors interpolation provides a solution that is constant around each node, on each subinterval defined by two consecutive midpoints of the node set. To calculate the required interpolants, we call the generic `scipy.interpolate.interp1d` function with the `kind='nearest'` option. It produces an instance of the `_Interpolator1D` class with only the evaluation method available.

The following example shows its result on a simple trigonometric function $f(x) = \sin(3*x)$ on the interval from 0 to 1:

```
In [1]: import numpy as np, matplotlib.pyplot as plt; \
   ...: from scipy.interpolate import interp1d
In [2]: nodes = np.linspace(0, 1, 5); \
   ...: print nodes
[ 0.    0.25  0.5   0.75  1.  ]
In [3]: def f(t): return np.sin(3 * t)
```

```
In [4]: x = np.linspace(0, 1, 100)          # the domain
In [5]: interpolant = interp1d(nodes, f(nodes), kind='nearest')
In [6]: plt.rc('text', usetex=True)
   ...: plt.figure(); \
   ...: plt.axes().set_aspect('equal'); \
   ...: plt.plot(nodes, f(nodes), 'ro', label='Nodes'); \
   ...: plt.plot(x, f(x), 'r-', label=r'f(x)=\sin(3x)'); \
   ...: plt.plot(x, interpolant(x), 'b--', label='Interpolation'); \
   ...: plt.title("Nearest-neighbor approximation"); \
   ...: plt.ylim(-0.05, 1.05); \
   ...: plt.xlim(-0.5, 1.05); \
   ...: plt.show()
```

This produces the following graph:

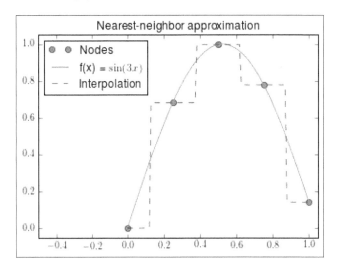

Lagrange interpolation

In Lagrange interpolation, we seek a polynomial that agrees with a target function at the set of nodes. In the `scipy.interpolate` module, we have three ways to solve this problem:

- The `BarycentricInterpolator` subclass of `_Interpolator1D` implements a very stable algorithm based upon approximation by rational functions. This class has an evaluation method, plus two methods to add/update nodes of the fly: `.add_xi` and `.set_yi`.

- A procedural scheme `barycentric_interpolate` is syntactic sugar for the previous class, with the evaluation method applied on a prescribed domain.

- A numerically unstable functional scheme, `lagrange`, computes a `numpy.poly1d` instance of the interpolating polynomial. If the nodes are few and wisely chosen, this method allows us to deal with derivative, integration, and root-solving problems associated with the target function, somewhat reliably.

Let us experiment with this interpolation mode on the infamous `Runge` example: Find an interpolation polynomial for the function $f(x) = 1/(1+x^2)$ in the interval from -5 to 5, with two sets of equally distributed nodes:

```
In [7]: from scipy.interpolate import BarycentricInterpolator, \
   ...: barycentric_interpolate, lagrange
In [8]: nodes = np.linspace(-5, 5, 11); \
   ...: x = np.linspace(-5,5,1000); \
   ...: print nodes
[-5. -4. -3. -2. -1.  0.  1.  2.  3.  4.  5.]
In [9]: def f(t): return 1. / (1. + t**2)
In [10]: interpolant = BarycentricInterpolator(nodes, f(nodes))
In [11]: plt.figure(); \
   ....: plt.subplot(121, aspect='auto'); \
   ....: plt.plot(x, interpolant(x), 'b--', \
   ....:             label="Lagrange Interpolation"); \
   ....: plt.plot(nodes, f(nodes), 'ro', label='nodes'); \
   ....: plt.plot(x, f(x), 'r-', label="original"); \
   ....: plt.legend(loc=9); \
   ....: plt.title("11 equally distributed nodes")
Out[11]: <matplotlib.text.Text at 0x10a5fbe50>
```

The `BarycentricInterpolator` class allows adding the extra nodes and updating interpolant in an optimal way, without the need to recalculate from scratch:

```
In [12]: newnodes = np.linspace(-4.5, 4.5, 10); \
   ....: print newnodes
[-4.5 -3.5 -2.5 -1.5 -0.5  0.5  1.5  2.5  3.5  4.5]
In [13]: interpolant.add_xi(newnodes, f(newnodes))
In [14]: plt.subplot(122, aspect='auto'); \
   ....: plt.plot(x, interpolant(x), 'b--', \
```

```
....:                   label="Lagrange Interpolation"); \
....: plt.plot(nodes, f(nodes), 'ro', label='nodes'); \
....: plt.plot(x, f(x), 'r-', label="original"); \
....: plt.legend(loc=8); \
....: plt.title("21 equally spaced nodes"); \
....: plt.show()
```

We obtain the following results:

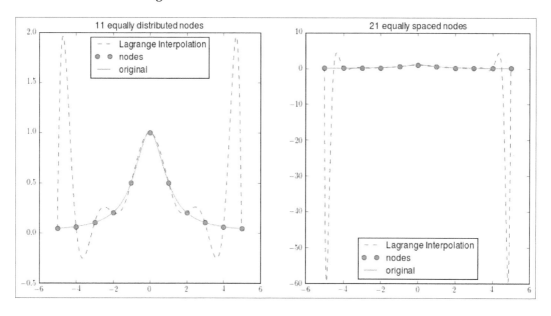

The Runge example shows one of the shortcomings of very simple interpolation. Although the interpolant accurately approximates the function in the interior of the interval, it shows a very large deviation at the endpoints.

The same methods to initialize our interpolants can be called to request information about them. The following short session illustrates this point:

```
In [15]: print interpolant.xi
[-5.  -4.  -3.  -2.  -1.   0.   1.   2.   3.   4.   5.  -4.5 -3.5
 -2.5 -1.5 -0.5  0.5  1.5  2.5  3.5  4.5]
In [16]: print interpolant.yi.squeeze()
[ 0.04  0.06  0.1   0.2   0.5   1.    0.5   0.2   0.1   0.06  0.04
  0.05  0.08  0.14  0.31  0.8   0.8   0.31  0.14  0.08  0.05]
```

The procedural scheme has a simpler syntax, but lacks the flexibility to update nodes on the fly:

```
In [17]: y = barycentric_interpolate(nodes, f(nodes), x)
```

The functional scheme also enjoys a simple syntax:

```
In [18]: g = lagrange(nodes, f(nodes)); \
   ....: print g
             10               9             8             7             6
3.858e-05 x   + 6.268e-19 x - 0.002149 x + 3.207e-17 x + 0.04109 x
             5             4             3             2
 + 5.117e-17 x - 0.3302 x - 2.88e-16 x + 1.291 x - 1.804e-16 x
```

Hermite interpolation

The goal of `Hermite` interpolation is the computation of a polynomial that agrees with a target function and some of its derivatives in a finite set of nodes. We accomplish this task numerically with two schemes:

- A subclass of `_Interpolator1DWithDerivatives`, `KroghInterpolator`, which has the `.derivative` method to compute a representation for any derivative of the interpolant and the `.derivatives` method to evaluate it.

- A `krogh_interpolate` function, which is syntactic sugar for the previous class with the evaluation method applied on a prescribed domain.

Let us showcase these routines with Bernstein's example: Compute the Hermite interpolation to the absolute value function in the interval from `-1` to `1` with ten equally distributed nodes, providing one derivative on each node.

 The nodes need to be fed in an increasing order. For every node in which we present derivatives, we repeat the node as many times as necessary. For each occurrence of a node in `xi`, we place on `yi` the evaluation of the function and its derivatives, at the same entry levels.

```
In [19]: from scipy.interpolate import KroghInterpolator
In [20]: nodes = np.linspace(-1, 1, 10); \
   ....: x = np.linspace(-1, 1, 1000)
In [21]: np.set_printoptions(precision=3, suppress=True)
In [22]: xi = np.repeat(nodes, 2); \
   ....: print xi; \
```

```
    ....: yi = np.ravel(np.dstack((np.abs(nodes), np.sign(nodes)))); \
    ....: print yi
[-1.     -1.     -0.778 -0.778 -0.556 -0.556 -0.333 -0.333 -0.111
 -0.111  0.111  0.111  0.333  0.333  0.556  0.556  0.778  0.778
  1.      1.    ]
[ 1.     -1.      0.778 -1.      0.556 -1.      0.333 -1.      0.111
 -1.      0.111  1.      0.333  1.      0.556  1.      0.778  1.
  1.      1.    ]
In [23]: interpolant = KroghInterpolator(xi, yi)
In [24]: plt.figure(); \
    ....: plt.axes().set_aspect('equal'); \
    ....: plt.plot(x, interpolant(x), 'b--', \
    ....:          label='Hermite Interpolation'); \
    ....: plt.plot(nodes, np.abs(nodes), 'ro'); \
    ....: plt.plot(x, np.abs(x), 'r-', label='original'); \
    ....: plt.legend(loc=9); \
    ....: plt.title('Bernstein example'); \
    ....: plt.show()
```

This gives the following diagram:

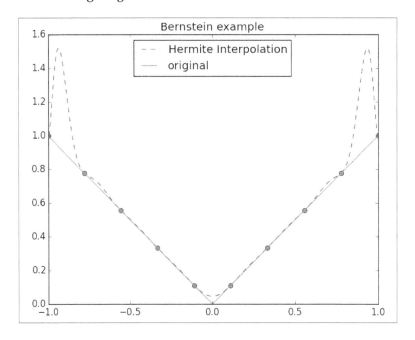

Piecewise polynomial interpolation

By prescribing the degrees of several polynomials and a finite set of nodes, we can construct an interpolator that has on each subinterval between two consecutive nodes, a polynomial arc with the required order. We can construct interpolants of this characteristic with the following procedures:

- A subclass of `_Interpolator1DWithDerivatives`, the `PiecewisePolynomial` class, with methods for evaluating interpolants and its derivatives, or appending new nodes

- For the special case of piecewise linear interpolation, the `interp1d` utility creates an instance of the `_Interpolator1D` class with only the evaluation method

- A `piecewise_polynomial_interpolate` function, which is syntactic sugar for the `PiecewisePolynomial` class, with the evaluation method applied on a prescribed domain

Let us revisit the first example in this section. First, we try piecewise linear interpolation with `interp1d`. Second, we apply piecewise quadratic interpolation (all pieces have order 2) with the correct derivatives at every node by using `PiecewisePolynomial`.

```
In [25]: from scipy.interpolate import PiecewisePolynomial
In [26]: nodes = np.linspace(0, 1, 5); \
   ....: x = np.linspace(0, 1, 100)
In [27]: def f(t): return np.sin(3 * t)
In [28]: interpolant = interp1d(nodes, f(nodes), kind='linear')
In [29]: plt.figure(); \
   ....: plt.subplot(121, aspect='equal'); \
   ....: plt.plot(x, interpolant(x), 'b--', label="interpolation"); \
   ....: plt.plot(nodes, f(nodes), 'ro'); \
   ....: plt.plot(x, f(x), 'r-', label="original"); \
   ....: plt.legend(loc=8); \
   ....: plt.title("Piecewise Linear Interpolation")
Out[29]: <matplotlib.text.Text at 0x107be0390>
In [30]: yi = np.zeros((len(nodes), 2)); \
   ....: yi[:,0] = f(nodes); \
   ....: yi[:,1] = 3 * np.cos(3 * nodes); \
   ....: print yi
[[ 0.     3.    ]
```

```
    [ 0.682   2.195]
    [ 0.997   0.212]
    [ 0.778 -1.885]
    [ 0.141 -2.97 ]]
In [31]: interpolant = PiecewisePolynomial(nodes, yi, orders=2)
In [32]: plt.subplot(122, aspect='equal'); \
   ....: plt.plot(x, interpolant(x), 'b--', label="interpolation"); \
   ....: plt.plot(nodes, f(nodes), 'ro'); \
   ....: plt.plot(x, f(x), 'r-', label="original"); \
   ....: plt.legend(loc=8); \
   ....: plt.title("Piecewise Quadratic interpolation"); \
   ....: plt.show()
```

This gives the following diagram:

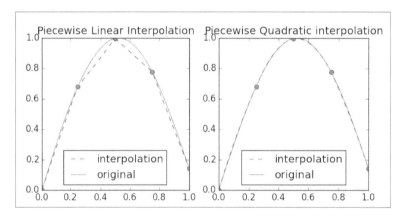

In this image, the piecewise quadratic interpolation and the original function are virtually indistinguishable. We need to go to the computation of the absolute values of differences (of the function together with its first and second derivatives) to actually realize the error of computation. The following is a crude computation that approximates these errors and illustrates the use of the .derivatives method:

```
In [33]: np.abs(f(x) - interpolant(x)).max()
Out[33]: 0.0093371930045896279
In [34]: f1prime = lambda t: 3 * np.cos(3 * t); \
   ....: np.abs(f1prime(x) - interpolant.derivatives(x)).max()
Out[34]: 10.589218385920123
In [35]: f2prime = lambda t: -9 * np.sin(3 * x); \
   ....: np.abs(f2prime(x) - interpolant.derivatives(x,der=2)).max()
Out[35]: 9.9980773091170505
```

A great advantage of piecewise polynomial approximation is the flexibility of using polynomials of different degrees on different subintervals. For instance, with the same set of nodes, we can use lines in the first and last subintervals and cubics for the others:

```
In [36]: interpolant = PiecewisePolynomial(nodes, yi, \
    ....:                                  orders=[1,3,3,1])
```

The other great advantage of the implementation of this interpolation scheme is the ease with which we may add new nodes, without the need to recalculate from scratch. For example, to add a new node after the last one, we issue:

```
In [37]: interpolant.append(1.25, np.array([f(1.25)]))
```

Spline interpolation

Univariate splines are a special case of piecewise polynomials. They possess a high degree of smoothness at places where the polynomial pieces connect. These functions can be written as linear combinations of basic splines with minimal support with respect to a given degree, smoothness, and set of nodes.

Univariate spline interpolations using `splines` of the order of up to five may be carried out by the `interp1d` function with the appropriate `kind` option. This function creates instances of `_Interpolator1DWithDerivatives`, with the corresponding class methods. The computations are performed through calls to routines in the Fortran library `FITPACK`. The following example shows the different possibilities:

```
In [38]: splines = ['zero', 'slinear', 'quadratic', 'cubic', 4, 5]; \
    ....: g = KroghInterpolator([0,0,0,1,1,1,2,2,2,3,3,3], \
    ....:                        [10,0,0,1,0,0,0.25,0,0,0.125,0,0]); \
    ....: f = lambda t: np.log1p(g(t)); \
    ....: x = np.linspace(0,3,100); \
    ....: nodes = np.linspace(0,3,11)
In [39]: plt.figure()
In [40]: for k in xrange(6):
    ....:     interpolant = interp1d(nodes, f(nodes), \
    ....:                            kind = splines[k])
    ....:     plt.subplot(2,3,k+1, aspect='equal')
    ....:     plt.plot(nodes, f(nodes), 'ro')
    ....:     plt.plot(x, f(x), 'r-', label='original')
    ....:     plt.plot(x, interpolant(x), 'b--', \
```

```
    ....:                    label='interpolation')
    ....:         plt.title('{0} spline'.format(splines[k]))
    ....:         plt.legend()
In [41]: plt.show()
```

This gives the following diagram:

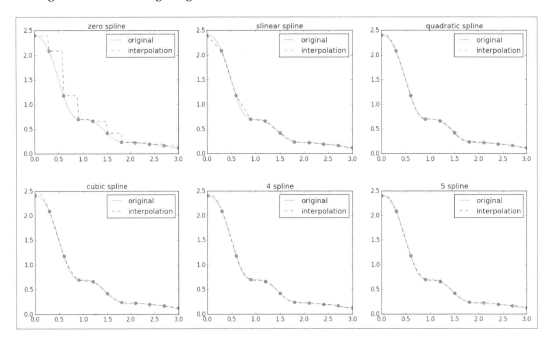

The zero spline is very similar to the nearest-neighbors approximation, although in this case the interpolant is constant between each choice of two consecutive nodes. The slinear spline is exactly the same as the piecewise linear interpolation. However, the algorithm that performs this interpolation through splines is slower.

For any given problem setup, there are many different possible spline interpolations with the same degrees, nodes, and evaluations. The output also depends on the position and the number of knots, for example. Unfortunately, the `interp1d` function only allows the control of nodes and values; the algorithm uses the simplest possible settings in terms of knot computation.

Note for instance that the cubic spline interpolation in the previous example does not preserve the monotonicity of the target function. It is possible to force the monotonicity of the interpolant in this case by carefully imposing restrictions on derivatives or the location of knots. We have a special function that achieves this task for us by using **piecewise monotonic cubic Hermite interpolation (PCHIP)**, implemented through the Fritsch-Carlson algorithm. This simple algorithm is carried out either by the `PchipInterpolator` subclass of `_Interpolator1DWithDerivatives`, or through its equivalent procedural function, `pchip_interpolate`.

```
In [42]: from scipy.interpolate import PchipInterpolator
In [43]: interpolant = PchipInterpolator(nodes, f(nodes))
In [44]: plt.figure(); \
    ....: plt.axes().set_aspect('equal'); \
    ....: plt.plot(nodes, f(nodes), 'ro'); \
    ....: plt.plot(x, f(x), 'r-', label='original'); \
    ....: plt.plot(x, interpolant(x), 'b--', label='interpolation'); \
    ....: plt.title('PCHIP interpolation'); \
    ....: plt.legend(); \
    ....: plt.show()
```

This gives the following graph:

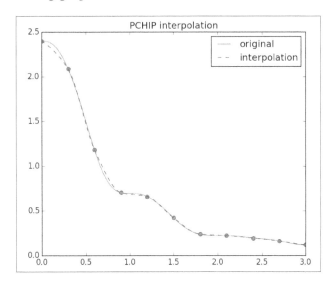

Generic spline interpolation, where we have actual control over all the different parameters that affect the quality of splines, is handled by the InterpolatedUnivariateSpline class. In this case, all computations are carried out by wrappers of routines from the Fortran library FITPACK. It is possible to access these wrappers in a procedural way, through a set of functions in the scipy. interpolate module. The following table shows a match between class methods, the corresponding procedural functions, and the FITPACK routines that they call:

Operation	Object-oriented implementation	Procedural	FITPACK
Instantiation of interpolant	InterpolatedUnivariateSpline	splrep	CURFIT
Reporting knots of spline	object.get_knots()	splrep	
Reporting spline coefficients	object.get_coeffs()	splrep	CURFIT
Evaluation of spline	object()	splev	SPLEV
Derivative	object.derivative()	splder	
Evaluation of derivatives	object.derivatives()	splev, spalde	SPLDER, SPALDE
Antiderivative	object.antiderivative()	splantider	
Definite integral	object.integral()	splint	SPLINT
Roots (for cubic splines)	object.roots()	sproot	SPROOT

 The values obtained by the.get_coeffs method are the coefficients of the spline as a linear combination of B-splines.

Let us show how to approximate the area under the graph of the target function by computing the integral of the corresponding interpolation spline of order 5.

```
In [45]: from scipy.interpolate import InterpolatedUnivariateSpline \
   ....: as IUS
In [46]: interpolant = IUS(nodes, f(nodes), k=5)
In [47]: area = interpolant.integral(0,3); \
   ....: print area
2.14931665485
```

Multivariate interpolation

Bivariate interpolation with splines can be performed with `interp2d` in the `scipy.interpolate` module. This is a very simple class that allows only the evaluation method and has three basic spline interpolation modes coded: linear, cubic, and quintic. It offers no control over knots or weights. To create a representation of the bivariate spline, the `interp2d` function calls the Fortran routine `SURFIT` from the library `FITPACK` (which, sadly, is not actually meant to perform interpolation!). To evaluate the spline numerically, the module calls the routine `BISPEV`.

Let us show by example the usage of `interp2d`. We first construct an interesting bivariate function to interpolate a random choice of 100 nodes on its domain and present a visualization:

```
In [1]: import numpy as np, matplotlib.pyplot as plt; \
   ...: from mpl_toolkits.mplot3d.axes3d import Axes3D
In [2]: def f(x, y): return np.sin(x) + np.sin(y)
In [3]: t = np.linspace(-3, 3, 100); \
   ...: domain = np.meshgrid(t, t); \
   ...: X, Y = domain; \
   ...: Z = f(*domain)
In [4]: fig = plt.figure(); \
   ...: ax1 = plt.subplot2grid((2,2), (0,0), aspect='equal'); \
   ...: p = ax1.pcolor(X, Y, Z); \
   ...: fig.colorbar(p); \
   ...: CP = ax1.contour(X, Y, Z, colors='k'); \
   ...: ax1.clabel(CP); \
   ...: ax1.set_title('Contour plot')
In [5]: nodes = 6 * np.random.rand(100, 2) - 3; \
   ...: xi = nodes[:, 0]; \
   ...: yi = nodes[:, 1]; \
   ...: zi = f(xi, yi)
In [6]: ax2 = plt.subplot2grid((2,2), (0,1), aspect='equal'); \
   ...: p2 = ax2.pcolor(X, Y, Z); \
   ...: ax2.scatter(xi, yi, 25, zi) ; \
   ...: ax2.set_xlim(-3, 3); \
   ...: ax2.set_ylim(-3, 3); \
   ...: ax2.set_title('Node selection')
```

```
In [7]: ax3 = plt.subplot2grid((2,2), (1,0), projection='3d', \
   ...:                         colspan=2, rowspan=2); \
   ...: ax3.plot_surface(X, Y, Z, alpha=0.25); \
   ...: ax3.scatter(xi, yi, zi, s=25); \
   ...: cset = ax3.contour(X, Y, Z, zdir='z', offset=-4); \
   ...: cset = ax3.contour(X, Y, Z, zdir='x', offset=-5); \
   ...: ax3.set_xlim3d(-5, 3); \
   ...: ax3.set_ylim3d(-3, 5); \
   ...: ax3.set_zlim3d(-4, 2); \
   ...: ax3.set_title('Surface plot')
In [8]: fig.tight_layout(); \
   ...: plt.show()
```

We obtain the following diagram:

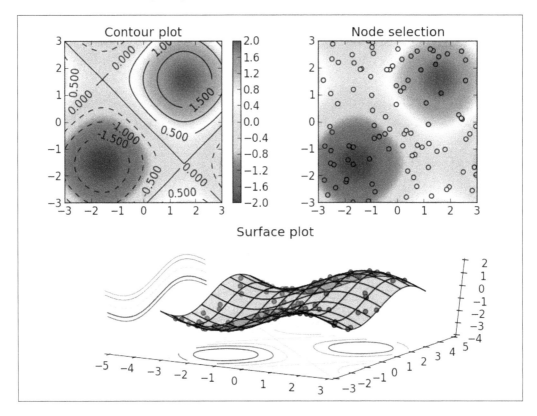

Piecewise linear interpolation with these nodes can be then performed as follows:

```
In [9]: from scipy.interpolate import interp2d
In [10]: interpolant = interp2d(xi, yi, zi, kind='linear')
In [11]: plt.figure(); \
    ....: plt.axes().set_aspect('equal'); \
    ....: plt.pcolor(X, Y, interpolant(t, t)); \
    ....: plt.scatter(xi, yi, 25, zi); \
    ....: CP = plt.contour(X, Y, interpolant(t, t), colors='k'); \
    ....: plt.clabel(CP); \
    ....: plt.xlim(-3, 3); \
    ....: plt.ylim(-3, 3); \
    ....: plt.title('Piecewise linear interpolation'); \
    ....: plt.show()
```

In spite of its name (`interp2d`), this process is not an actual interpolation but a crude approximation that tries to fit the data. In general, each time you run this code, you will obtain results of different quality. Luckily, this is not the case with the rest of the interpolation routines that follow!

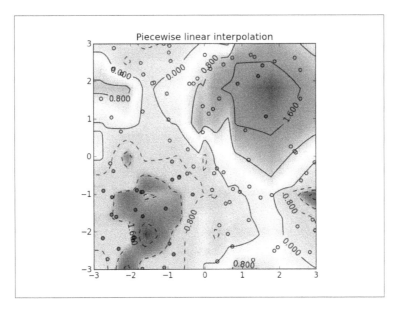

If the location of the nodes is not optimal, we are likely to obtain a warning:

```
Warning:     No more knots can be added because the
number of B-spline coefficients
    already exceeds the number of data points m. Probably
causes: either
    s or m too small. (fp>s)
  kx,ky=1,1 nx,ny=11,14 m=100 fp=0.002836 s=0.000000
```

Note that in the previous example, the evaluation of the interpolant is performed with a call to two one-dimensional arrays. In general, to evaluate an interpolant g computed with `interp2d` on a rectangular grid that can be realized as the Cartesian product of two one-dimensional arrays (`tx` of size m and `ty` of size n), we issue `g(tx, ty)`; this gives us a two-dimensional array of size *m* x *n*.

The quality of the result is, of course, deeply linked to the density and structure of the nodes. Increasing their number or imposing their location on a rectangular grid improves matters. In the case of nodes forming a rectangular grid, an actual interpolation, with a much faster and accurate method is accomplished by means of the `RectBivariateSpline` class. This function is a wrapper to the Fortran routine `REGRID` in the `FITPACK` library.

Let us now choose 100 nodes on a rectangular grid and recalculate, as follows:

```
In [12]: ti = np.linspace(-3, 3, 10); \
   ....: xi, yi = np.meshgrid(ti, ti); \
   ....: zi = f(xi, yi)
In [13]: from scipy.interpolate import RectBivariateSpline
In [14]: interpolant = RectBivariateSpline(ti, ti, zi, kx=1, ky=1)
In [15]: plt.figure(); \
   ....: plt.axes().set_aspect('equal'); \
   ....: plt.pcolor(X, Y, interpolant(t, t)); \
   ....: CP = plt.contour(X, Y, interpolant(t, t), colors='k'); \
   ....: plt.clabel(CP); \
   ....: plt.scatter(xi, yi, 25, zi); \
   ....: plt.xlim(-3, 3); \
   ....: plt.ylim(-3, 3); \
```

```
....: plt.title('Piecewise linear interpolation, \
....: rectangular grid'); \
....: plt.show()
```

 As with the case of `interp2d`, to evaluate an interpolant `g` computed with `RectBivariateSpline` on a rectangular grid that can be realized as the Cartesian product of two one-dimensional arrays (`tx` of size `m` and `ty` of size `n`), we issue `g(tx, ty)`; this gives us a two-dimensional array of size m x n.

This gives an actual interpolation now:

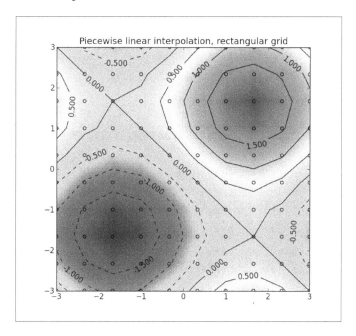

The volume integral under the graph is very accurate (the actual integral of the target function in the given domain is zero):

```
In [16]: interpolant.integral(-3, 3, -3, 3)
Out[16]: 2.636779683484747e-16
```

Let us examine some of the different pieces of information that we receive from this class, in this case:

- The degrees of the interpolant:

```
In [17]: interpolant.degrees
Out[17]: (1, 1)
```

- The sum of the squared residuals of the spline approximation returned:

```
In [18]: interpolant.fp
Out[18]: 0.0
In [19]: interpolant.get_residual()
Out[19]: 0.0
```

- The coefficients of the interpolant:

```
In [20]: np.set_printoptions(precision=5, suppress=True)
In [21]: print interpolant.get_coeffs()
[-0.28224 -0.86421 -1.13653 -0.98259 -0.46831  0.18607
  0.70035  0.85429  0.58197  0.       -0.86421 -1.44617
 -1.71849 -1.56456 -1.05028 -0.39589  0.11839  0.27232
  0.       -0.58197 -1.13653 -1.71849 -1.99082 -1.83688
 -1.3226  -0.66821 -0.15394  0.       -0.27232 -0.85429
 -0.98259 -1.56456 -1.83688 -1.68294 -1.16867 -0.51428
  0.        0.15394 -0.11839 -0.70035 -0.46831 -1.05028
 -1.3226  -1.16867 -0.65439 -0.        0.51428  0.66821
  0.39589 -0.18607  0.18607 -0.39589 -0.66821 -0.51428
 -0.        0.65439  1.16867  1.3226   1.05028  0.46831
  0.70035  0.11839 -0.15394  0.        0.51428  1.16867
  1.68294  1.83688  1.56456  0.98259  0.85429  0.27232
  0.        0.15394  0.66821  1.3226   1.83688  1.99082
  1.71849  1.13653  0.58197  0.       -0.27232 -0.11839
  0.39589  1.05028  1.56456  1.71849  1.44617  0.86421
  0.       -0.58197 -0.85429 -0.70035 -0.18607  0.46831
  0.98259  1.13653  0.86421  0.28224]
```

- The location of the knots:

```
In [22]: interpolant.get_knots()
(array([-3. , -3. , -2.33333, -1.66667, -1. , -0.33333,
         0.33333,  1. ,  1.66667,  2.33333,  3. ,  3. ]),
 array([-3. , -3. , -2.33333, -1.66667, -1. , -0.33333,
         0.33333,  1. ,  1.66667,  2.33333,  3. ,  3. ]))
```

Smoother results can be obtained with piecewise cubic splines. In the previous example, we can accomplish this task by setting kx = 3 and ky = 3:

```
In [23]: interpolant = RectBivariateSpline(ti, ti, zi, kx=3, ky=3)
In [24]: fig = plt.figure(); \
   ....: ax = fig.add_subplot(121, projection='3d',aspect='equal'); \
   ....: ax.plot_surface(X, Y, interpolant(t, t), alpha=0.25, \
   ....:                  rstride=5, cstride=5); \
   ....: ax.scatter(xi, yi, zi, s=25); \
   ....: C = ax.contour(X, Y, interpolant(t, t), zdir='z', \
   ....:                  offset=-4); \
   ....: C = ax.contour(X, Y, interpolant(t, t), zdir='x',\
   ....:                  offset=-5); \
   ....: ax.set_xlim3d(-5, 3); \
   ....: ax.set_ylim3d(-3, 5); \
   ....: ax.set_zlim3d(-4, 2); \
   ....: ax.set_title('Cubic interpolation, RectBivariateSpline')
```

Among all possible interpolations with cubic splines, there is a special case that minimizes the curvature. We have one implementation of this particular case, by means of a clever iterative algorithm that converges to the solution. It relies on the following three key concepts:

- Delaunay triangulations of the domain using the nodes as vertices
- Bezier cubic polynomials supported on each triangle using a Cough-Tocher scheme
- Estimation and imposition of gradients to minimize curvature

An implementation is available through the CloughTocher2dInterpolator function in the scipy.interpolate module, or through the black-box function griddata in the same module with the method='cubic' option. Let us compare the outputs:

```
In [25]: from scipy.interpolate import CloughTocher2DInterpolator
In [26]: nodes = np.dstack((np.ravel(xi), np.ravel(yi))).squeeze(); \
   ....: zi = f(nodes[:, 0], nodes[:, 1])
In [27]: interpolant = CloughTocher2DInterpolator(nodes, zi)
In [28]: ax = fig.add_subplot(122, projection='3d', aspect='equal'); \
   ....: ax.plot_surface(X, Y, interpolant(X, Y), alpha=0.25, \
   ....:                  rstride=5, cstride=5); \
   ....: ax.scatter(xi, yi, zi, s=25); \
```

```
....: C = ax.contour(X, Y, interpolant(X, Y), zdir='z', \
....:                 offset=-4); \
....: C = ax.contour(X, Y, interpolant(X, Y), zdir='x', \
....:                 offset=-5); \
....: ax.set_xlim3d(-5, 3); \
....: ax.set_ylim3d(-3, 5); \
....: ax.set_zlim3d(-4, 2); \
....: ax.set_title('Cubic interpolation, \
....: CloughTocher2DInterpolator'); \
....: plt.show()
```

 Unlike the cases of `interp2d` and `RectBivariateSpline`, to evaluate an interpolant g computed with `CloughTocher2DInterpolator` on a rectangular grid `X, Y = domain`, we issue `g(X, Y)` or `g(*domain)`.

This gives the following diagram:

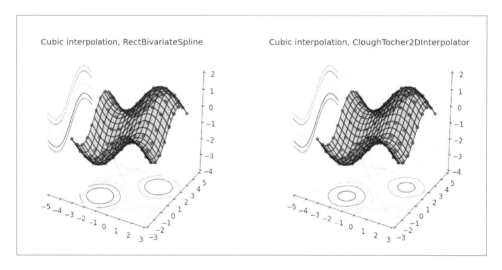

The black-box procedural function called `griddata` also allows us to access piecewise linear interpolation in multiple dimensions, as well as multidimensional nearest-neighbors interpolation.

```
In [29]: from scipy.interpolate import griddata
In [30]: Z = griddata(nodes, zi, (X, Y), method='nearest')
In [31]: plt.figure(); \
```

```
....: plt.axes().set_aspect('equal'); \
....: plt.pcolor(X, Y, Z); \
....: plt.colorbar(); \
....: plt.title('Nearest-neighbors'); \
....: plt.show()
```

This gives us the following not-too-impressive diagram:

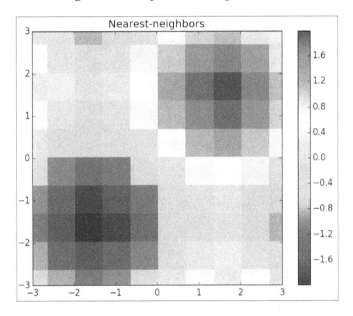

There is one more interpolation mode to consider: radial basis function interpolation. The aim here is to interpolate with a linear combination of radial functions of the form `fk(x,y) = g(sqrt((x-xk)**2 + (y-yk)**2))`, centered at the points `(xk, yk)`, for the same function `g`. We may choose among seven standard functions `g` (listed below), or even choose our own:

- `'multiquadric': g(r) = sqrt((r/self.epsilon)**2 + 1)`
- `'inverse': g(r) = 1.0/sqrt((r/self.epsilon)**2 + 1)`
- `'gaussian': g(r) = exp(-(r/self.epsilon)**2)`
- `'linear': g(r) = r`
- `'cubic': g(r) = r**3`
- `'quintic': g(r) = r**5`
- `'thin_plate': g(r) = r**2 * log(r)`

The implementation is performed through the Rbf class. It can be initialized as usual with nodes and their evaluations. We also need to include the choice of radial function, and if necessary, the value of the epsilon parameter affecting the size of the bumps:

Let us run a couple of interpolations: first, by means of radial Gaussians with standard deviation epsilon = 2.0, and then, with a radial function based on sinc. Let us also go back to random nodes:

```
In [32]: from scipy.interpolate import Rbf
In [33]: nodes = 6 * np.random.rand(100, 2) - 3; \
   ....: xi = nodes[:, 0]; \
   ....: yi = nodes[:, 1]; \
   ....: zi = f(xi, yi)
In [34]: interpolant = Rbf(xi, yi, zi, function='gaussian', \
   ....:                        epsilon=2.0)
In [35]: plt.figure(); \
   ....: plt.subplot(121, aspect='equal'); \
   ....: plt.pcolor(X, Y, interpolant(X, Y)); \
   ....: plt.scatter(xi, yi, 25, zi); \
   ....: plt.xlim(-3, 3); \
   ....: plt.ylim(-3, 3)
Out[35]: (-3, 3)
In [36]: interpolant = Rbf(xi, yi, zi, function = np.sinc)
In [37]: plt.subplot(122, aspect='equal'); \
   ....: plt.pcolor(X, Y, interpolant(X, Y)); \
   ....: plt.scatter(xi, yi, 25, zi); \
   ....: plt.xlim(-3, 3); \
   ....: plt.ylim(-3, 3); \
   ....: plt.show()
```

This gives two very accurate interpolations, in spite of the unstructured nodes:

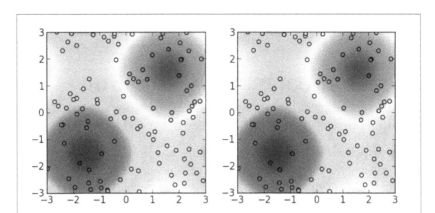

The last case for consideration is bivariate spline interpolation over a rectangular grid on a sphere. We obtain this interpolation with the RectSphereBivariateSpline function, which instantiates a subclass of SphereBivariateSpline through calls to the FITPACK routines SPGRID (to compute the representation of the spline) and BISPEV (for evaluation).

The implementation and evaluation are reminiscent of the Fortran coding methodology:

- To compute the spline representation, we issue the RectSphereBivariateSpline(u, v, data) command, where both u and v are one-dimensional arrays of strictly increasing positive values representing the angles (in radians) for the latitudes and longitudes (respectively) of the locations of the nodes.

- At evaluation time, if we require a two-dimensional representation of the interpolant on a refined grid of size m x n, we issue object(theta, phi) where both theta and phi are one-dimensional and strictly increasing, and must be contained in the domains defined by u and v above. The output is (in spite of what your documentation says) an m x n array.

Least squares approximation

Numerically, it is relatively simple to state the approximation problem for the least squares norm. This is the topic of this section.

Linear least squares approximation

In the context of linear least squares approximation, it is always possible to reduce the problem to solving a system of linear equations, as the following example shows:

Consider the sine function $f(x) = sin(x)$ in the interval from 0 to 1. We choose as approximants the polynomials of second degree: $\{a_0 + a_1 x + a_2 x^2\}$. To compute the values $[a_0, a_1, a_2]$ that minimize this problem, we first form a 3×3 matrix containing the pairwise dot products (the integral of the product of two functions) of the basic functions $\{1, x, x^2\}$ in the given interval. Because of the nature of this problem, we obtain a Hilbert matrix of order 3:

```
[    < 1, 1 >      < 1, x >     < 1, x^2 > ]      [  1    1/2  1/3 ]
[    < x, 1 >      < x, x >     < x, x^2 > ]  =   [ 1/2   1/3  1/4 ]
[ < x^2, 1 >    < x^2, x >    < x^2, x^2 > ]      [ 1/3   1/4  1/5 ]
```

The right-hand side of the system is the column vector with the dot product of the sine function with each basic function in the given interval:

```
[   < sin(x), 1 > ]        [              1 - cos(1) ]
[   < sin(x), x > ]   =    [         sin(1) - cos(1) ]
[ < sin(x), x^2 > ]        [ 2*sin(1) + cos(1)  - 2 ]
```

We compute the coefficients and the corresponding approximation polynomial as follows:

```
In [1]: import numpy as np, scipy.linalg as spla, \
   ...: matplotlib.pyplot as plt
In [2]: A = spla.hilbert(3); \
   ...: b = np.array([1-np.cos(1), np.sin(1)-np.cos(1), \
   ...:              2*np.sin(1)+ np.cos(1)-2])
In [3]: spla.solve(A, b)
Out[3]: array([-0.00746493,  1.09129978, -0.2354618 ])
In [4]: poly1 = np.poly1d(spla.solve(A, b)[::-1]); \
   ...: print poly1
        2
-0.2355 x + 1.091 x - 0.007465
```

In general, to resolve a linear least squares approximation problem for a basis with r elements, we need to solve a basic system of linear equations with r equations and r indeterminates. In spite of its apparent simplicity, this method is far from perfect. The two main reasons are the following:

- The system may be ill-conditioned, as was the case in the previous example.

- There is nonpermanence of the coefficients. The value of the coefficients depends very heavily on r. Increasing the dimension of the problem results in a new set of coefficients, different from the previous set.

There are ways to remediate the ill-conditioning of the system. One standard procedure is to construct an orthogonal basis from the original with the Gram-Schmidth and modified Gram-Schmidt orthogonalization methods. This topic is beyond the scope of this monograph, but a good reference for these methods can be read in *Chapter 1* of the book *Numerical Analysis* by Walter Gautschi, Birkhäuser, 1997.

A basis that always provides simple linear systems is the B-splines. All the systems involved are tridiagonal and thus are easily solvable without the need of complex operations. The object-oriented system coded in the `scipy.interpolate` module allows us to perform all these computations internally. This is a brief enumeration of the classes and subclasses involved:

- `UnivariateSpline` for splines in one dimension, or splines of curves in any dimension. We seldom use this class directly and resort instead to the `LSQUnivariateSpline` subclass.

- `BivariateSpline` for splines representing surfaces over nodes placed on a rectangle. As its univariate counterpart, this class must not be used directly. Instead, we utilize the `LSQBivariateSpline` subclass.

- `SphereBivariateSpline` for splines representing surfaces over nodes placed on a sphere. The computations must be carried out through the `LSQSphereBivariateSpline` subclass instead.

In all three cases, the base classes and their methods are their counterparts in the problem of interpolation. Refer to the *Interpolation* section for more information.

Let us illustrate these object-oriented techniques with a few selected examples:

Approximate the same `sine` function on the same domain, with cubic splines (`k` = 3), in the sense of least squares. First, note that we must provide a bounding box, a set of knots on the domain, and the weights `w` for the least squares approximation. We are also allowed to provide an optional smoothness parameter `s`. If `s` = 0, we obtain interpolation instead, and for large values of `s`, we achieve different degrees of smoothness of the resulting spline. To obtain a reliable (weighted) least squares approximation, a good choice is `s` = `len(w)` (which is what the routine does by default). Note also how small the computed error is:

```
In [5]: f = np.sin; \
   ...: x = np.linspace(0,1,100); \
   ...: knots = np.linspace(0,1,7)[1:-1]; \
   ...: weights = np.ones_like(x)
In [6]: from scipy.interpolate import LSQUnivariateSpline
In [7]: approximant = LSQUnivariateSpline(x, f(x), knots, k=3, \
   ...:                                    w = weights, bbox = [0, 1])
In [8]: spla.norm(f(x) - approximant(x))
Out[8]: 3.370175009262551e-06
```

A more convenient way to compute this error of approximation is to use the `.get_residual` method, as follows:

```
In [9]: approximant.get_residual()**(.5)
Out[9]: 3.37017500928446e-06
```

Approximate the two-dimensional function `sin(x)+sin(y)` on the `[-3,3]` x `[-3,3]` domain. We first choose a representation of the domain, a set of 100 suitable knots on a grid, and the set of weights. Since all inputs to the `LSQBivariateSpline` function must be one-dimensional arrays, we perform the corresponding conversions prior to calling the approximation function:

```
In [10]: def f(x, y): return np.sin(x) + np.sin(y); \
    ....: t = np.linspace(-3, 3, 100); \
    ....: domain = np.meshgrid(t, t); \
    ....: X, Y = domain; \
    ....: Z = f(*domain)
In [11]: X = X.ravel(); \
    ....: Y = Y.ravel(); \
    ....: Z = Z.ravel()
```

```
In [12]: kx = np.linspace(-3,3,12)[1:-1]; \
    ....: ky = kx.copy(); \
    ....: weights = np.ones_like(Z);
In [13]: from scipy.interpolate import LSQBivariateSpline
In [14]: approximant = LSQBivariateSpline(X, Y, Z, kx, kx, \
    ....:                                 w = weights)
In [15]: approximant.get_residual()
Out[15]: 0.0
```

It is also possible to perform this computation with the
RectBivariateSpline function. To achieve least squares
interpolation, we provide nodes (instead of knots, since these
will be computed automatically), weights w, and a smoothness
parameter s that is sufficiently large. A good choice is s =
len(w).

Nonlinear least squares approximation

In the context of nonlinear least square approximations, we do not usually have
the luxury of simple matrix representations. Instead, we make use of two variations
of an iterative process, the Levenberg-Marquardt algorithm, which is hosted in
the scipy.optimize module. The two versions, which correspond to the LMDER
and LMDIF routines from the Fortran library MINPACK, can be called through the
leastsq wrapper.

The following table lists all the options to this function:

Option	Description
func	error function F(a)
x0	starting estimate for the minimization, of size r
args	extra arguments to func, as a tuple
Dfun	function representing the Jacobian matrix of func
full_output	Boolean
col_deriv	Boolean
ftol	relative error desired in the sums of squares
xtol	relative error desired in the approximate solution
gtol	orthogonality desired between func and the columns of Dfun
maxfev	maximum number of calls. If zero, the number of calls is 100*(r+1)

Option	Description
epsfcn	if Dfun=None, we may specify a floating-point value as the step in the forward-difference approximation of the Jacobian matrix
factor	floating-point value between 0.1 and 100, indicating the initial step bound
diag	scale factors for each of the variables

The first variant of the algorithm is used when we have a trusted Jacobian for the error function. If this is not provided, a second variant of the algorithm is used, which approximates the Jacobian by forward differences. We illustrate both variants with several examples.

Let us start revisiting a previous example with this method, in order to see the differences in usage and accuracy. We will focus the computations on a partition of the interval from 0 to 1, with 100 uniformly spaced points:

```
In [16]: from scipy.optimize import leastsq
In [17]: def error_function(a):
   ....:     return a[0] + a[1] * x + a[2] * x**2 - np.sin(x)
In [18]: def jacobian(a): return np.array([np.ones(100), x, x**2])
In [19]: coeffs, success = leastsq(error_function, np.zeros((3,)))
In [20]: poly2 = np.poly1d(coeffs[::-1]); print poly2
        2
-0.2354 x + 1.091 x - 0.007232
In [21]: coeffs, success = leastsq(error_function, np.zeros((3,)), \
   ....:                           Dfun = jacobian, col_deriv=True)
In [22]: poly3 = np.poly1d(coeffs[::-1]); \
   ...:   print poly3
        2
-0.2354 x + 1.091 x - 0.007232
In [23]: map(lambda f: spla.norm(np.sin(x) - f(x)), \
   ....:     [poly1, poly2, poly3])
Out[23]:
[0.028112146265269783, 0.02808377541388009, 0.02808377541388009]
```

There is another function in the scipy.optimize module to perform nonlinear least squares approximation: curve_fit. It uses the same algorithm, but instead of an error function, we feed it a generic approximant g[a](x), together with a suitable domain for the independent variable x, and the output of the target function f on the same domain. We do need to input an initial estimate as well. The output is, together with the required coefficients, an estimation of the covariance of the said coefficients.

```
In [23]: from scipy.optimize import curve_fit
In [24]: def approximant(t, a, b, c):
    ....:     return a + b*t + c*t**2
In [25]: curve_fit(approximant, x, np.sin(x), \
    ....:          np.ones((3,)))
(array([-0.007232  ,  1.09078356, -0.23537796]),
 array([[  7.03274163e-07,  -2.79884256e-06,
           2.32064835e-06],
        [ -2.79884256e-06,   1.50223585e-05,
          -1.40659702e-05],
        [  2.32064835e-06,  -1.40659702e-05,
           1.40659703e-05]]))
```

In this section, we focus on the leastsq function exclusively. The goals and coding of both the functions are the same, but leastsq offers a more informative output on demand and more control over the different parameters of the Levenberg-Marquardt algorithm.

Let us experiment now with a few actual nonlinear problems:

In the first example, we will approximate the tan(2*x) function in the interval from 0 to 1 with rational functions where each of the polynomials has an at most degree of 1:

```
In [26]: def error_function(a):
    ....:     return (a[0] + a[1]*x)/(a[2] + a[3]*x) - np.tan(2*x)
In [27]: def jacobian(a):
    ....:     numerator = a[0] + a[1]*x
    ....:     denominator = a[2] + a[3]*x
    ....:     return np.array( [ 1./denominator, x/denominator, \
    ....:                       -1.0*numerator/denominator**2, \
    ....:                       -1.0*x*numerator/denominator**2 ])
```

To show the dependence of the initial estimation, we are going to experiment with three different choices: one that makes no sense (all zero coefficients), another that is a blind standard choice (with all entries equal to one), and the other choice that acknowledges the fact that the `tan(2*x)` function has a vertical asymptote. We will pretend that we do not know the exact location and approximate it to `0.78`. Our third initial estimation then represents a simple rational function with an asymptote at `0.78`.

A wrong initial estimate does not give us anything useful, obviously:

```
In [28]: x1 = np.zeros((4,)); \
   ....: x2 = np.ones((4,)); \
   ....: x3 = np.array([1,0,0.78,-1])
In [29]: coeffs, success = leastsq(error_function, x1); \
   ....: numerator = np.poly1d(coeffs[1::-1]); \
   ....: denominator = np.poly1d(coeffs[:1:-1]); \
   ....: print numerator, denominator

0

0

In [30]: coeffs, success = leastsq(error_function, x1, \
   ....:                            Dfun=jacobian, col_deriv=True); \
   ....: numerator = np.poly1d(coeffs[1::-1]); \
   ....: denominator = np.poly1d(coeffs[:1:-1]); \
   ....: print numerator, denominator

0

0
```

None of these two approximations using `x2` as the initial guess are satisfactory: the corresponding errors are huge, and neither solution has an asymptote in the interval from 0 to 1.

```
In [31]: coeffs, success = leastsq(error_function, x2); \
   ....: numerator = np.poly1d(coeffs[1::-1]); \
   ....: denominator = np.poly1d(coeffs[:1:-1]); \
   ....: print numerator, denominator; \
   ....: spla.norm(np.tan(2*x) - numerator(x) / denominator(x))

-9.729 x + 4.28

-1.293 x + 1.986

Out[31]: 220.59056436054016
```

```
In [32]: coeffs, success = leastsq(error_function, x2, \
    ....:                            Dfun=jacobian, col_deriv=True); \
    ....: numerator = np.poly1d(coeffs[1::-1]); \
    ....: denominator = np.poly1d(coeffs[:1:-1]); \
    ....: print numerator, denominator; \
    ....: spla.norm(np.tan(2*x) - numerator(x) / denominator(x))
-655.9 x + 288.5
-87.05 x + 133.8
Out[32]: 220.590564978504
```

The approximations using x3 as the initial guess are closer to the target function, and both of them have an acceptable asymptote.

```
In [33]: coeffs, success = leastsq(error_function, x3); \
    ....: numerator = np.poly1d(coeffs[1::-1]); \
    ....: denominator = np.poly1d(coeffs[:1:-1]); \
    ....: print numerator, denominator; \
    ....: spla.norm(np.tan(2*x) - numerator(x) / denominator(x))
0.01553 x + 0.02421
-0.07285 x + 0.05721
Out[33]: 2.185984698129936
In [34]: coeffs, success = leastsq(error_function, x3, \
    ....:                            Dfun=jacobian, col_deriv=True); \
    ....: numerator = np.poly1d(coeffs[1::-1]); \
    ....: denominator = np.poly1d(coeffs[:1:-1]); \
    ....: print numerator, denominator; \
    ....: spla.norm(np.tan(2*x) - numerator(x) / denominator(x))
17.17 x + 26.76
-80.52 x + 63.24
Out[34]: 2.1859846981334954
```

We can do much better, of course, but these simple examples will suffice for now.

If we desire to output more information to monitor the quality of approximation, we may do so with the full_output option set to True:

```
In [35]: approximation_info = leastsq(error_function, x3, \
    ....:                             full_output=True)
```

```
In [36]: coeffs = approximation_info[0]; \
   ....: print coeffs
[ 0.02420694  0.01553346  0.0572128  -0.07284579]
In [37]: message = approximation_info[-2]; \
   ....: print message
Both actual and predicted relative reductions in the sum of squares
  are at most 0.000000
In [38]: infodict = approximation_info[2]; \
   ....: print 'The algorithm performed \
   ....: {0:2d} iterations'.format(infodict['nfev'])
The algorithm performed 97 iterations
```

Although technically, the `leastsq` algorithm deals mostly with approximation to univariate functions, it is possible to work on multivariate functions with the aid of indices, raveling, unpacking (with the special * operator), and stable sums.

The usual sum of the floating-point numbers of ndarray with the numpy instance method sum (or with the numpy function sum) is far from stable. We firmly advise against using it for fairly large sums of numbers. The following example shows an undesired scenario, in which we try to add 4000 values:

```
>>> arr=np.array([1,1e20,1,-1e20]*1000,dtype=np.float64)
>>> arr.sum()     # The answer should be, of course, 2000
0.0
```

To resolve this situation, we make use of stable sums. In the math module, there is an implementation of the Shewchuk algorithm for this very purpose:

```
>>> from math import fsum
>>> fsum(arr)
2000.0
```

For more information about the Shewchuk algorithm, as well as other common pitfalls to avoid in scientific computing with floating-point arithmetic, we recommend the excellent guide *What Every Computer Scientist Should Know about Floating Point Arithmetic*, by David Goldberg. ACM Computing Surveys, 1991. vol. 23, pp. 5-48.

This process is best explained with an example. We start by generating the target function: an image of size 32×32 containing white noise on top of the addition of three spherical Gaussian functions with different locations, variances, and heights. We collect all these values in a 3×4 array that we name values. The first and second columns contain the x and y values of the coordinates of the centers. The third column contains the heights, and the fourth column contains the variances.

```
In [39]: def sphericalGaussian(x0, y0, h, v):
   ....:         return lambda x,y: h*np.exp(-0.5*((x-x0)**2+(y-y0)**2)/v)
   ....:
In [40]: domain = np.indices((32, 32)); \
   ....: values = np.random.randn(3,4); \
   ....: values[:,:2] += np.random.randint(1, 32, size=(3, 2)); \
   ....: values[:,2] += np.random.randint(1, 64, size=3); \
   ....: values[:,3] += np.random.randint(1, 16, size=3); \
   ....: print values
[[ 17.43247918   17.15301326   34.86691265    7.84836966]
 [  5.5450271    20.68753512   34.41364835    4.78337552]
 [ 24.44909459   27.28360852    0.62186068    9.15251106]]
In [41]: img = np.random.randn(32,32)
In [42]: for k in xrange(3):
   ....:         img += sphericalGaussian(*values[k])(*domain)
```

Let us assume that we do not know the centers, heights, and variances, and wish to estimate them from the target image img. We then create an error function to compute the 12 coefficients that are packed in the 3×4 array a. Note the role of the numpy function ravel and the instance method reshape in ensuring that the data is handled correctly:

```
In [43]: from math import fsum
In [44]: def error_function(a):
   ....:         a = a.reshape(3,4)
   ....:         cx = a[:,0]      # x-coords
   ....:         cy = a[:,1]      # y-coords
   ....:         H = a[:,2]       # heights
   ....:         V = a[:,3]       # variances
   ....:         guess = np.zeros_like(img)
   ....:         for i in xrange(guess.shape[0]):
```

```
....:             for j in xrange(guess.shape[1]):
....:                 arr = H*np.exp(-0.5*((i-cx)**2+(j-cy)**2)/V)
....:                 guess[i,j] = fsum(arr)
....:         return np.ravel(guess-img)
```

Starting the process of least squares in this situation with guarantees of success requires a somewhat close initial guess. For this particular example, we are going to produce our initial guess from the array called `values`:

```
In [45]: x0 = np.vectorize(int)(values); \
   ....: print x0
[[17 17 34  7]
 [ 5 20 34  4]
 [24 27  0  9]]
In [46]: leastsq(error_function, x0)
Out[46]:
(array([ 17.43346907,  17.14219682,  34.82077187,   7.85849653,
          5.52511918,  20.68319748,  34.28559808,   4.8010449 ,
         25.19824918,  24.02286107,   3.87170006,   0.5289382 ]),
 1)
```

Let us now visually compare both the target image `img` and its minimization by the following procedure:

```
In [47]: coeffs, success = _; \
   ....: coeffs = coeffs.reshape(3,4)
In [48]: output = np.zeros_like(img)
In [49]: for k in xrange(3):
   ....:     output += sphericalGaussian(*coeffs[k])(*domain)
In [50]: plt.figure(); \
   ....: plt.subplot(121); \
   ....: plt.imshow(img); \
   ....: plt.title('target image'); \
   ....: plt.subplot(122); \
   ....: plt.imshow(output); \
   ....: plt.title('computed approximation'); \
   ....: plt.show()
```

This gives the following diagram:

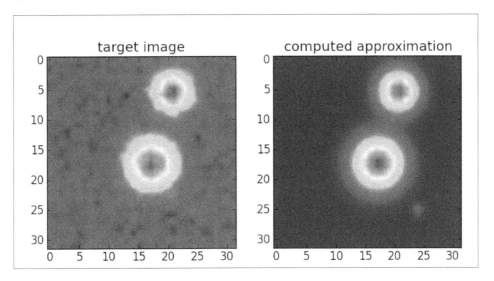

Summary

In this chapter, we have explored two basic problems in the field of approximation theory: interpolation and approximation in the sense of least squares. We learned that there are three different modes to approach solutions to these problems in SciPy:

- A procedural mode, that offers quick numerical solutions in the form of `ndarrays`.
- A functional mode that offers `numpy` functions as the output.
- An object-oriented mode, with great flexibility through different classes and their methods. We use this mode when we require from our solutions extra information (such as information about roots, coefficients, knots, and errors), or related objects (such as the representation of derivatives or antiderivatives).

We explored in detail all the different implementations for the interpolation coded in the `scipy.interpolate` module, and learned in particular that those related to splines are wrappers of several routines in the Fortran library FITPACK.

In the case of linear approximations in the least squares sense, we learned that we may achieve solutions either through systems of linear equations (by means of techniques from the previous chapter), or in the case of spline approximation, through wrappers to Fortran routines in the FITPACK library. All of these functions are coded in the `scipy.interpolate` module.

For nonlinear approximation in the least squares sense, we found two variants of the Levenberg-Marquardt iterative algorithm coded in the `scipy.optimize` module. These are in turn calls to the Fortran routines LMDER and LMDIF from the library MINPACK.

In the next chapter, we will master techniques and applications of differentiation and integration.

3

Differentiation and Integration

In this chapter, we will master some classical and state-of-the-arts techniques to perform the two core operations in Calculus (and, by extension, in Physics and every engineering field): differentiation and integration of functions.

Motivation

Common to the design of railway or road building (especially for highway exits), as well as those crazy loops in many roller coasters, is the solution of differential equations in two or three dimensions that address the effect of curvature and centripetal acceleration on moving bodies. In the 1970s, Werner Stengel studied and applied several models to attack this problem and, among the many solutions he found, one struck as particularly brilliant—the employment of clothoid loops (based on sections of Cornu's spiral). The first looping coaster designed with this paradigm was constructed in 1976 in the Baja Ridge area of Six Flags Magic Mountain, in Valencia, California, USA. It was coined the Great American Revolution, and it featured the very first vertical loop (together with two corkscrews, for a total of three inversions).

The tricky part of the design was based on a system of differential equations, whose solution depended on the integration of Fresnel-type sine and cosine integrals, and then selecting the appropriate sections of the resulting curve. Let's see the computation and plot of these interesting functions:

```
In [1]: import numpy as np, matplotlib.pyplot as plt; \
   ...: from scipy.special import fresnel
In [2]: np.info(fresnel)
fresnel(x[, out1, out2])

(ssa,cca)=fresnel(z) returns the Fresnel sin and cos integrals:
integral(sin(pi/2 * t**2),t=0..z) and
integral(cos(pi/2 * t**2),t=0..z)
for real or complex z.
In [3]: ssa, cca = fresnel(np.linspace(-4, 4, 1000))
In [4]: plt.plot(ssa, cca, 'b-'); \
   ...: plt.axes().set_aspect('equal'); \
   ...: plt.show()
```

This results in the following plot:

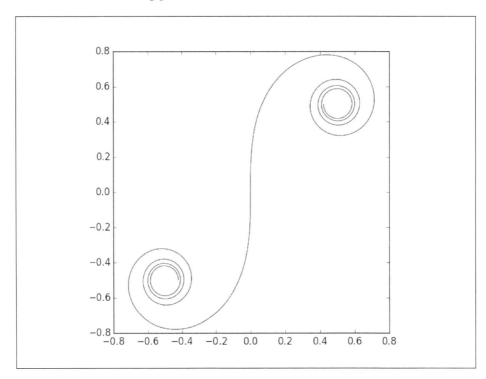

The importance of Fresnel integrals granted them a permanent place in SciPy libraries. There are many other useful integrals that shared the same fate, and now lie ready for action in the module `scipy.special`. For a complete list of all those integrals, as well as implementation of other relevant functions and their roots or derivatives, refer to the online documentation of `scipy.special` at `http://docs.scipy.org/doc/scipy-0.13.0/reference/special.html`, or in *Chapter 4, SciPy for Numerical Analysis* of Francisco Blanco-Silva's *Learning SciPy for Numerical and Scientific Computing*.

For all the other functions that did not make it to this ample list, we still need robust solutions to the computation of their roots, derivatives, or integrals. In this chapter, we will focus on computational devices that allow the last two operations.

 The calculation of (or approximation to the) roots of any given function will be covered in the next chapter.

Differentiation

There are three ways to approach the computation of derivatives:

- Numerical differentiation refers to the process of approximation of the derivative of a given function at a point. In SciPy, we have the following procedures, which will be covered in detail:

    - For generic univariate functions, the central difference formula with fixed spacing.

    - It is always possible to perform numerical differentiation via Cauchy's theorem, which transforms the derivative into a definite integral. This integral is then treated with the techniques of numerical integration explained in the upcoming section.

- Symbolic differentiation refers to computation of functional expressions of derivatives of functions, pretty much in the same way that we would do manually. It is termed symbolic because unlike its numerical counterpart, symbols take the role of variables, rather than numbers or vectors of numbers. To perform symbolic differentiation, we require a **computer algebra system (CAS)**, and in the SciPy stack, this is achieved mainly through the `SymPy` library (see `http://docs.sympy.org/latest/index.html`). Symbolic differentiation and posterior evaluation is a good option as a substitute of numerical differentiation for very basic functions. However, in general, this method leads to overcomplicated and inefficient code. The speed of purely numerical differentiation is preferred, in spite of the possible occurrence of errors.

- Automatic differentiation is another set of techniques to numerically evaluate the derivative of a function. It is not based upon any approximation schema. This is without a doubt the most powerful option in the context of high derivatives of multivariate functions.

> In the SciPy stack, this is performed through different unrelated libraries. Some of the most reliable are Theano (http://deeplearning.net/software/theano/) or FuncDesigner (http://www.openopt.org/FuncDesigner). For a comprehensive description and analysis of these techniques, a very good resource can be found at http://alexey.radul.name/ideas/2013/introduction-to-automatic-differentiation/.

Numerical differentiation

The most basic scheme for numerical differentiation is performed with the central difference formula with uniformly spaced nodes. To maintain symmetry, an odd number of nodes is required to guarantee smaller roundoff errors. An implementation of this simple algorithm is available in the module scipy.misc.

> For information about the module scipy.misc, and enumeration of its basic routines, refer to the online documentation at http://docs.scipy.org/doc/scipy-0.13.0/reference/misc.html.

To approximate the first and second derivatives of the polynomial function, for example, $f(x) = x^5$ at $x=1$ with 15 equally spaced nodes (centered at $x=1$) at distance dx=1e-6, we could issue the following command:

```
In [1]: import numpy as np
In [2]: from scipy.misc import derivative
In [3]: def f(x): return x**5
In [4]: derivative(f, 1.0, dx=1e-6, order=15)
Out[4]: 4.9999999997262723
In [5]: derivative(f, 1.0, dx=1e-6, order=15, n=2)
Out[5]: 19.998683310705456
```

Somewhat accurate, yet still disappointing since the actual values are 5 and 20, respectively.

> Another flaw of this method (at least with respect to the implementation coded in SciPy) is the fact that the result relies on possibly large sums, and these are not stable. As users, we could improve matters by modifying the loop in the source of scipy.misc.derivative with the Shewchuk algorithm, for instance.

Symbolic differentiation

Exact differentiation for polynomials can be achieved through the module `numpy.polynomial`:

```
In [6]: p = np.poly1d([1,0,0,0,0,0]); \
   ...: print p
   5
1 x
```

```
In [7]: np.polyder(p,1)(1.0)        In [7]: p.deriv()(1.0)
Out[7]: 5.0                          Out[7]: 5.0
In [8]: np.polyder(p,2)(1.0)        In [8]: p.deriv(2)(1.0)
Out[8]: 20.0                         Out[8]: 20.0
```

Symbolic differentiation is another way to achieve exact results:

```
In [9]: from sympy import diff, symbols
In [10]: x = symbols('x', real=True)
In [11]: diff(x**5, x)              In [12]: diff(x**5, x, x)
Out[11]: 5*x**4                      Out[12]: 20*x**3
In [13]: diff(x**5, x).subs(x, 1.0)
Out[13]: 5.00000000000000
In [14]: diff(x**5, x, x).subs(x, 1.0)
Out[14]: 20.0000000000000
```

Note the slight improvement (both in notation and simplicity of coding) when we differentiate more involved functions than simple polynomials. For example, for $g(x) = e^{-x}sinx$ at $x=1$:

```
In [15]: def g(x): return np.exp(-x) * np.sin(x)
In [16]: derivative(g, 1.0, dx=1e-6, order=101)
Out[16]: -0.11079376536871781
In [17]: from sympy import sin as Sin, exp as Exp
In [18]: diff(Exp(-x) * Sin(x), x).subs(x, 1.0)
Out[18]: -0.110793765306699
```

A great advantage of symbolic differentiation over its numerical or automatic counterparts is the possibility to compute partial derivatives with extreme ease. Let's illustrate this point by calculating a fourth partial derivative of the multivariate function $h(x,y,z) = e^{xyz}$ at $x=1$, $y=1$, and $z=2$:

```
In [19]: y, z = symbols('y z', real=True)
In [20]: diff(Exp(x * y * z), z, z, y, x).subs({x:1.0, y:1.0, z:2.0})
Out[20]: 133.003009780752
```

Automatic differentiation

The third method employs automatic differentiation. For this example, we will use the library `FuncDesigner`:

```
In [21]: from FuncDesigner import oovar, exp as EXP, sin as SIN
In [22]: X = oovar('X'); \
    ....: G = EXP(-X) * SIN(X)
In [23]: G.D({X: 1.0}, X)
Out[23]: -0.11079376530669924
```

The result is obviously more accurate than the one obtained with numerical differentiation. Also, there was no need to provide any extra parameters.

Integration

To achieve a definite integration of functions on suitable domains, we have mainly two methods—**Numerical integration** and **Symbolic integration**.

Numerical integration refers to the approximation of a definite integral by a quadrature process. Depending on how the function *f(x)* is given, the domain of integration, the knowledge of its singularities, and the choice of quadrature, we have different ways to attack this problem:

- For univariate polynomials, exact integration is achieved algebraically on each finite interval

- For functions given as a finite set of samples over their domain:
    - The composite trapezoidal rule
    - Simpson's trapezoidal rules
    - Romberg integration scheme

- For generic univariate functions given as Python functions, on finite intervals:
    - Fixed-order Gaussian quadrature
    - Fixed-tolerance Gaussian quadrature
    - Simple non-adaptive quadrature, by applying 21-, 43- and 87-point Gauss-Kronron rules
    - Simple adaptive quadrature, by subdivision and quadrature on each subinterval

- A blind global adaptive quadrature based on the 21-point Gauss-Kronrod quadrature within each subinterval, with an acceleration process (the Peter Wynn's epsilon algorithm):

    ◦ A global adaptive quadrature based on the previous, but with a user-provided location of singularities/discontinuities

    ◦ An adaptive Romberg integration scheme

- For univariate functions given as Python functions on unbounded intervals, there is a global adaptive quadrature. The process transforms the infinite interval into a semi-open interval, and applies a 15-point Gauss-Kronrod quadrature within each subinterval.

- For multivariate functions given as Python functions on type I domains (which will be described shortly), a method that applies adaptive univariate quadratures, iteratively on each dimension is generally used.

In many cases, it is also possible to perform exact integration, even for not-bounded domains, with the aid of symbolic computation. In the SciPy stack, to this effect, we have an implementation of the Risch algorithm for elementary functions, and Meijer G-functions for non-elementary integrals. Both methods are housed in the SymPy libraries. Unfortunately, these symbolic procedures do not work for all functions. And due to the complexity of the generated codes, in general, the solutions obtained by this method are by no means as fast as any numerical approximation.

Symbolic integration

The definite integral of a polynomial function on a finite domain [a,b] can be computed very accurately via the Fundamental Theorem of Calculus, through the module numpy.polynomial. For instance, to calculate the integral of the polynomial $p(x)=x^5$ on the interval [-1,1], we could issue:

```
In [1]: import numpy as np
In [2]: p = np.poly1d([1,0,0,0,0,0]); \
   ...: print p; \
   ...: print p.integ()
   5
1 x
        6
0.1667 x
In [3]: p.integ()(1.0) - p.integ()(-1.0)
Out[3]: 0.0
```

In general, obtaining exact values for a definite integral of a generic function is hard and computationally inefficient. This is possible in some cases through symbolic integration, with the aid of the Risch algorithm (for elementary functions) and Meijer G-functions (for non-elementary integrals). Both methods can be called with the common routine integrate in the library SymPy. The routine is clever enough to decide which algorithm to use, depending on the source function.

Let's show you a few examples starting with the definite integral of the polynomial from the previous case:

```
In [4]: from sympy import integrate, symbols
In [5]: x, y = symbols('x y', real=True)
In [6]: integrate(x**5, x)
Out[6]: x**6/6
In [7]: integrate(x**5, (x, -1, 1))
Out[7]: 0
```

Let's try something more complicated. The definite integral of the function $f(x) = e^{-x}sinx$ on the interval [0,1]:

```
In [8]: from sympy import N, exp as Exp, sin as Sin
In [9]: integrate(Exp(-x) * Sin(x), x)
Out[9]: -exp(-x)*sin(x)/2 - exp(-x)*cos(x)/2
In [10]: integrate(Exp(-x) * Sin(x), (x, 0, 1))
Out[10]: -exp(-1)*sin(1)/2 - exp(-1)*cos(1)/2 + 1/2
In [11]: N(_)
Out[11]: 0.245837007000237
```

Symbolic integration, when it works, treats singularities the right way:

```
In [12]: integrate(Sin(x) / x, x)
Out[12]: Si(x)
In [13]: integrate(Sin(x) / x, (x, 0, 1))
Out[13]: Si(1)
In [14]: N(_)
Out[14]: 0.946083070367183
In [15]: integrate(x**(1/x), (x, 0, 1))
Out[15]: 1/2
```

Integration over unbounded domains is also possible:

```
In [16]: from sympy import oo
In [17]: integrate(Exp(-x**2), (x,0,+oo))
Out[17]: sqrt(pi)/2
```

It is even possible to perform multivariate integration:

```
In [18]: integrate(Exp(-x**2-y**2), (x, -oo, +oo), (y, -oo, +oo))
Out[18]: pi
```

However, we need to stress this point strongly—symbolic integration is not efficient (and might not work!) for simple cases, as the following example shows:

```
In [19]: integrate(Sin(x)**Sin(x), x)
Integral(sin(x)**sin(x), x)
In [20]: integrate(Sin(x)**Sin(x), (x, 0, 1))
Integral(sin(x)**sin(x), (x, 0, 1))
```

Even when it works for simple cases, it generates complicated code, and might use too many computational resources.

Numerical integration

The optimal way to address these problems is to obtain good approximations instead, with the aid of numerical integration. There are different techniques, according to the type of function and integration domain. Let's examine them in detail.

Functions without singularities on finite intervals

The basic problem in numerical integration is the approximation to the definite integral of any function $f(x)$ on a finite interval [a,b]. In general, if the function $f(x)$ does not have singularities or discontinuities, we can obtain easy quadrature formulas by integrating different interpolations with piecewise polynomials (since these are evaluated exactly):

- The composite trapezoidal rule is achieved by integration of a piecewise linear interpolator (every two consecutive nodes)

- Simpson's rule is achieved by integrating a piecewise polynomial interpolator, where every two consecutive subintervals we fit a parabola

- In the previous case, if we further impose Hermite interpolation, we obtain the composite Simpson's rule

We have efficient algorithms for composite trapezoidal and composite Simpson's rules in the module `scipy.integrate` through the routines `cumtrapz` and `simps`, respectively. Let's show you how to use these simple quadrature formulas for the polynomial example:

```
In [21]: from scipy.integrate import cumtrapz, simps
In [22]: def f(x): return x**5
In [23]: nodes = np.linspace(-1, 1, 100)
In [24]: simps(f(nodes), nodes)
Out[24]: -1.3877787807814457e-17
In [25]: cumtrapz(f(nodes), nodes)
Out[25]:
array([ -1.92221161e-02,  -3.65619927e-02,  -5.21700680e-02,
        -6.61875756e-02,  -7.87469280e-02,  -8.99720915e-02,
        -9.99789539e-02,  -1.08875683e-01,  -1.16763077e-01,
        -1.23734908e-01,  -1.29878257e-01,  -1.35273836e-01,
        -1.39996314e-01,  -1.44114617e-01,  -1.47692240e-01,
        -1.50787532e-01,  -1.53453988e-01,  -1.55740523e-01,
        -1.57691741e-01,  -1.59348197e-01,  -1.60746651e-01,
        -1.61920310e-01,  -1.62899066e-01,  -1.63709727e-01,
        -1.64376231e-01,  -1.64919865e-01,  -1.65359463e-01,
        -1.65711607e-01,  -1.65990811e-01,  -1.66209700e-01,
        -1.66379187e-01,  -1.66508627e-01,  -1.66605982e-01,
        -1.66677959e-01,  -1.66730153e-01,  -1.66767180e-01,
        -1.66792794e-01,  -1.66810003e-01,  -1.66821177e-01,
        -1.66828145e-01,  -1.66832283e-01,  -1.66834598e-01,
        -1.66835799e-01,  -1.66836364e-01,  -1.66836598e-01,
        -1.66836678e-01,  -1.66836700e-01,  -1.66836703e-01,
        -1.66836703e-01,  -1.66836703e-01,  -1.66836703e-01,
        -1.66836700e-01,  -1.66836678e-01,  -1.66836598e-01,
        -1.66836364e-01,  -1.66835799e-01,  -1.66834598e-01,
        -1.66832283e-01,  -1.66828145e-01,  -1.66821177e-01,
        -1.66810003e-01,  -1.66792794e-01,  -1.66767180e-01,
        -1.66730153e-01,  -1.66677959e-01,  -1.66605982e-01,
        -1.66508627e-01,  -1.66379187e-01,  -1.66209700e-01,
        -1.65990811e-01,  -1.65711607e-01,  -1.65359463e-01,
        -1.64919865e-01,  -1.64376231e-01,  -1.63709727e-01,
        -1.62899066e-01,  -1.61920310e-01,  -1.60746651e-01,
        -1.59348197e-01,  -1.57691741e-01,  -1.55740523e-01,
        -1.53453988e-01,  -1.50787532e-01,  -1.47692240e-01,
        -1.44114617e-01,  -1.39996314e-01,  -1.35273836e-01,
        -1.29878257e-01,  -1.23734908e-01,  -1.16763077e-01,
        -1.08875683e-01,  -9.99789539e-02,  -8.99720915e-02,
        -7.87469280e-02,  -6.61875756e-02,  -5.21700680e-02,
        -3.65619927e-02,  -1.92221161e-02,  -1.73472348e-17])
```

> The routine `cumtrapz` computes cumulative integrals over the designated subintervals. The last entry of the output is therefore the value of the quadrature we seek. We could, of course, report only the required integral by simply accessing that entry:
>
> ```
> In [26]: cumtrapz(f(nodes), nodes)[-1]
> Out[26]: -1.7347234759768071e-17
> ```

The implementation of these two algorithms does not compute the interpolators explicitly. The final formulas are the target here, and the way it is coded in SciPy is by means of Newton-Cotes quadratures.

The routines to perform Newton-Cotes are hidden (in the sense that they are not reported in the tutorials or documentation in the official pages of SciPy) and are meant to be used only internally by `cumtrapz` or `simps`. They provide only the corresponding coefficients that multiply the function evaluation at the nodes.

However, Newton-Cotes quadrature formulas are usually very accurate by themselves in the right scenarios. They can be used to compute better approximations in many cases, without being subjected to conform to trapezoidal or Simpson's rules.

Let's show you how it works for our running example, now with only four equally spaced nodes in the interval `[-1,1]`:

```
In [27]: from scipy.integrate import newton_cotes
In [28]: coefficients, abs_error = newton_cotes(3, equal=True); \
   ....: nodes = np.linspace(-1, 1, 4); \
   ....: print coefficients
[ 0.375   1.125   1.125   0.375]
In [29]: integral = (coefficients * f(nodes)).sum(); \
   ....: print integral
0.0
In [30]: from math import fsum
In [31]: integral = fsum(coefficients * f(nodes)); \
   ....: print integral
-7.8062556419e-18
```

If the nodes of our choice happen to be equally spaced, then there is an improvement of the trapezoidal rule in a special case—if the number of subintervals is a power of two. In that case, we may use the Romberg rule—an improvement that uses the Richardson extrapolation. We can access it with the routine `romb` in the same module.

Let's compare results with our running example, this time using 64 subintervals of size 1/32 in the interval [-1,1]:

```
In [32]: from scipy.integrate import romb
In [33]: nodes = np.linspace(-1, 1, 65)
In [34]: romb(f(nodes), dx=1./32)
0.0
```

We have the option to report the table that shows the Richardson extrapolation from the given nodes:

```
In [35]: romb(f(nodes), dx=1./32, show=True)
       Richardson Extrapolation Table for Romberg Integration
====================================================================
 0.00000
 0.00000  0.00000
 0.00000  0.00000  0.00000
 0.00000  0.00000  0.00000  0.00000
 0.00000  0.00000  0.00000  0.00000  0.00000
 0.00000  0.00000  0.00000  0.00000  0.00000  0.00000
====================================================================
Out[35]: 0.0
```

We might not have any preference for the choice of nodes, but still have our mind set in using Romberg's rule for our numerical integration scheme. In that case, we could use the routine romberg, for which we only need to provide the expression of a function and the limits of integration. Optionally, we can provide absolute or relative tolerances for the error (which are both set by default to 1.48e-8):

```
In [36]: from scipy.integrate import romberg
In [37]: romberg(f, -1, 1, show=True)
Romberg integration of <function vfunc at 0x10ffa8c08> from [-1, 1]
  Steps   StepSize   Results
      1   2.000000   0.000000
      2   1.000000   0.000000   0.000000
The final result is 0.0 after 3 function evaluations.
Out[37]: 0.0
```

Another possibility is to use Gaussian quadrature formulas. These are more powerful, since the accuracy of the approximations is gained through computing, internally, the best possible choice of nodes. There are two basic routines in the module `scipy.integrate` that perform implementations of this algorithm: `quadrature`, if we want to specify tolerance, or `fixed_quad`, if we wish to specify the number of nodes (but not their locations!):

```
In [38]: from scipy.integrate import quadrature, fixed_quad
In [39]: value, absolute_error = quadrature(f, -1, 1, tol=1.49e-8); \
   ....: print value
0.0
In [40]: value, absolute_error = fixed_quad(f, -1, 1, n=4); \
   ....: print value                                      # four nodes
-9.45424294407e-16
```

A more advanced method to perform Gaussian quadrature, using an adaptive scheme, is obtained through the function `quad` in the module `scipy.integrate`. This function is a wrapper of the routine QAGS in the Fortran library QUADPACK. The algorithm breaks the domain of integration into several subintervals and on each of them, performs a 21-point Gaussian-Kronrod quadrature rule. Further acceleration is achieved with Peter Wynn's epsilon algorithm.

 For more information on QAGS as well as the other routines in the QUADPACK libraries, refer to the netlib repositories: `http://www.netlib.org/quadpack/`.

Let's compare this with our running example:

```
In [41]: from scipy.integrate import quad
In [42]: value, absolute_error = quad(f, -1, 1); \
   ....: print value
0.0
```

We could obtain implementation details by setting the optional argument `full_output` to `True`. This gives us an additional Python dictionary with useful information:

```
In [43]: value, abs_error, info = quad(f, -1, 1, full_output=True)
In [44]: info.keys()
Out[44]: ['rlist', 'last', 'elist', 'iord', 'alist', 'blist',
          'neval']
In [45]: print "{0} function evaluations".format(info['neval'])
21 function evaluations
In [46]: print "Used {0} subintervals".format(info['last'])
Used 1 subintervals
```

To fully understand all the different outputs of `info`, we need to know about the underlying algorithm computing the Gaussian quadratures. These particular routines use the Clensaw-Curtis method, a technique based on Chebyshev moments.

In the preceding example, by default, the code tried to use 50 Chebyshev moments. Due to the simplicity of the integrand, and since only one subinterval was needed, it was necessary only to use one of those moments. When we report the 50-entry one-dimensional outputs `rlist`, `elist`, `alist`, and `blist` from the dictionary info, we can disregard the information offered by the last 49 entries of each of them:

```
In [47]: np.set_printoptions(precision=2, suppress=True)
In [48]: print info['rlist']  # integral approx on subintervals
[  0.00e+000    2.32e+077    6.93e-310    0.00e+000    0.00e+000
   0.00e+000    0.00e+000    0.00e+000    0.00e+000    0.00e+000
   0.00e+000    0.00e+000    0.00e+000    0.00e+000    0.00e+000
   0.00e+000    0.00e+000    0.00e+000    0.00e+000    0.00e+000
   0.00e+000    6.45e-314    2.19e-314    6.93e-310    0.00e+000
   0.00e+000    0.00e+000    0.00e+000    0.00e+000    0.00e+000
   0.00e+000    0.00e+000    0.00e+000    0.00e+000    0.00e+000
   0.00e+000    0.00e+000    0.00e+000    0.00e+000    0.00e+000
   0.00e+000    0.00e+000   -1.48e-224    2.19e-314    6.93e-310
   0.00e+000    0.00e+000    0.00e+000    0.00e+000    0.00e+000]
In [49]: print info['elist']  # abs error estimates on subintervals
[  3.70e-015    2.32e+077    3.41e-322    0.00e+000    0.00e+000
   0.00e+000    0.00e+000    0.00e+000    0.00e+000    0.00e+000
   0.00e+000    0.00e+000    0.00e+000    0.00e+000    0.00e+000
   0.00e+000    0.00e+000    0.00e+000    0.00e+000    7.30e+245
   2.19e-314    6.93e-310    0.00e+000    0.00e+000    0.00e+000
   4.74e+246    2.20e-314    6.93e-310    0.00e+000    0.00e+000
   0.00e+000    0.00e+000    0.00e+000    0.00e+000   -9.52e+207
   2.19e-314    6.93e-310    0.00e+000    0.00e+000    0.00e+000
   0.00e+000    0.00e+000    0.00e+000    0.00e+000    0.00e+000
   0.00e+000    2.00e+000    2.00e+000    2.27e-322    1.05e-319]
In [50]: print info['alist']   # subintervals left end points
[-1.  2.  0.  0.  0.  0.  0.  0.  0.  0.  0.  0.  0.  0.  0.
  0.  0.  0.  0.  0.  0.  0.  0.  0.  0.  0.  0.  0.  0.  0.
  0.  0.  0.  0.  0.  0.  0.  0.  0.  0.  0.  0.  0.  0.  0.
  0.  0.  0.  0.  0.]
In [51]: print info['blist']   # subintervals right end pts
[ 1.  2.  0.  0.  0.  0.  0.  0.  0.  0.  0.  0.  0.  0.  0.
  0.  0.  0.  0.  0.  0.  0.  0.  0.  0.  0.  0.  0.  0.  0.
  0.  0.  0.  0.  0.  0.  0.  0.  0.  0.  0.  0.  0.  0.  0.
  0.  0.  0.  2. -0.]
```

It is possible to impose a different number of Chebyshev moments to be used. We do so with the optional parameter `maxp1`, which imposes an upper bound to this number (rather than fixing it, for optimal results).

Oscillatory integrals of the form $f(x)cos(wx)$ or $f(x)sin(wx)$, even when $f(x)$ is smooth, are especially tricky. The integrator `quad` can tackle these integrals by calling the routine QAWO in QUADPACK. We can employ this method by specifying the arguments `weight='cos'` or `weight='sin'`, with `wvar=w`.

For example, for the integral of $g(x) = sin(x)e^x$ on the interval `[-10,10]`, we compare this method with a basic `quad`. We could do the following:

```
In [52]: def f(x): return np.sin(x) * np.exp(x)
In [53]: g = np.exp
In [54]: quad(g, -10, 10, weight='sin', wvar=1)
Out[54]: (3249.4589405744427, 5.365398805302381e-08)
In [55]: quad(f, -10, 10)
Out[55]: (3249.458940574436, 1.1767634585879705e-05)
```

Note the significant gain in absolute error.

> For details and the theory behind all the quadrature formulas that we have explored in this section, a good reference is *Chapter 3, Numerical Differentiation and Integration* of *Walter Gautchi's Numerical Analysis*.

Functions with singularities on bounded domains

The second case of integration is that of definite integrals on a finite interval `[a,b]` of functions with singularities. We contemplate two cases: weighted functions and generic functions with singularities.

Weighted functions

Weighted functions can be realized as products of the $f(x)w(x)$ kind for some smooth function $f(x)$ with a non-negative weight function $w(x)$ containing singularities. An illustrative example is given by $cos(\pi x/2)/\sqrt{x}$. We could regard this case as the product of $cos(\pi x/2)$ with $w(x)=1/\sqrt{x}$. The weight presents a single singularity at $x=0$, and is smooth otherwise.

The usual way to treat these integrals is by means of weighted Gaussian quadrature formulas. For example, to perform principal value integrals of functions of the form $f(x)/(x-c)$, we issue `quad` with the arguments `weight='cauchy'` and `wvar=c`. This calls the routine QAWC from QUADPACK.

Let's experiment with the Fresnel-type sine integral of $g(x) = sin(x)/x$ on the interval [-1,1] and compare it with `romberg`:

```
In [56]: value, abs_error = quad(f, -1, 1, weight='cauchy',wvar=0); \
   ....: print value
1.89216614073
In [57]: romberg(g, -1, 1)
Out[57]: 2.35040238729
```

In the case of integrals of functions with weights $(x-a)^{\alpha}(b-x)^{\beta}$, where a and b are the endpoints of the domain of integration and both `alpha` and `beta` are greater than -1, we issue `quad` with the arguments `weight='alg'` and `wvar=(alpha, beta)`. This calls the routine `QAWS` from `QUADPACK`.

Let's experiment with the Fresnel-type cosine integral of $g(x)=cos(\pi x/2)/\sqrt{x}$ on the interval [0,1] and compare it with `quadrature`:

```
In [58]: def f(x): return np.cos(np.pi * x * 0.5)
In [59]: def g(x): return np.cos(np.pi * x * 0.5) / np.sqrt(x)
In [60]: value, abs_error = quad(f, 0, 1, weight='alg', \
                                wvar=(-0.5,0)); \
   ....: print value
1.55978680075
In [61]: quadrature(g, 0, 1)
quadrature.py:178: AccuracyWarning: maxiter (50) exceeded. Latest
difference = 3.483182e-04
  AccuracyWarning)
Out[61]: (1.5425452942607543, 0.00034831815190772275)
```

If the weight has the form $w(x)=(x-a)^{\alpha}(b-x)^{\beta}ln(x-a)$, $w(x)=(x-a)^{\alpha}(b-x)^{\beta}ln(b-x)$, or $w(x)=(x-a)^{\alpha}(b-x)^{\beta}ln(x-a)ln(b-x)$, we issue `quad` with the arguments `weight='alg-loga'`, or `weight='alg-logb'` or `weight='alg-log'` respectively, and in each case, `wvar=(alpha, beta)`. For example, for the function $g(x)=x^2ln(x)$ on the interval [0,1], we could issue the following:

```
In [62]: def f(x): return x**2
In [63]: def g(x): return x**2 * np.log(x)
In [64]: value, abs_error = quad(f, 0, 1, weight='alg-loga', \
                                wvar=(0,0)); \
   ....: print value
-0.111111111111
```

The actual value of this integral is -1/9.

General functions with singularities

In general, we might be handling functions with singularities that do not conform to the nice forms *f(x)w(x)* that were indicated in the previous section. In that case, if we are aware of the locations of the singularities, we could indicate so to the integrator quad with the optional argument points. The integrator quad calls the routine QAGP in QUADPACK. For example, for the function *g(x) = floor(x)ln(x)* that observes a singularity on each integer number, to integrate on the interval [1,8], we could issue the following:

```
In [65]: def g(x): return np.floor(x) * np.log(x)
In [66]: quad(g, 1, 8, points=np.arange(8)+1)
Out[66]: (45.802300241541005, 5.085076830895319e-13)
```

Compare this to a simple quad computation without indicating any singularities, as the next lines of code show:

```
In [67]: quad(g, 1, 8)
quadpack.py:295: UserWarning: The maximum number of subdivisions (50)
has been achieved.
  If increasing the limit yields no improvement it is advised to
analyze the integrand in order to determine the difficulties.  If the
position of a local difficulty can be determined (singularity,
discontinuity) one will probably gain from splitting up the interval
and calling the integrator on the subranges.  Perhaps a special-
purpose integrator should be used.
  warnings.warn(msg)
Out[67]: (45.80231242134967, 8.09803045669355e-05)
```

Integration on unbounded domains

The versatile integrator quad is also able to compute definite integrals on unbounded domains using adaptive quadrature formulas, by means of a call to the routine QAGI from QUADPACK. This process does not work with 'cauchy', or any of the 'alg'-type weight options.

In general, if the functions to integrate do not present singularities, the approximations are reliable. The presence of singularities gives unreliable integrals, as the following example suggests:

```
In [68]: def f(x): return 2 * np.exp(-x**2) / np.sqrt(np.pi)
In [69]: value, absolute_error = quad(f, 0, np.inf); \
   ....: print value
1.0
In [70]: def f(x): return np.sin(x)/x
In [71]: integrate(Sin(x)/x, (x, 0, oo))
```

```
Out[71]: pi/2
In [72]: value, absolute_error = quad(f, 0, np.inf); \
   ....: print value                              # ouch!
2.24786796347
```

In the case of oscillatory integrals in unbounded domains, besides issuing `quad` with the argument `weight='cos'` or `weight='sin'` and the corresponding `wvar` parameter, we can also place an upper bound on the number of cycles to use internally. We do so by setting the optional argument `limlst` to the desired bound. It is usually a good idea to set it to something larger than three. For example, for the Fourier-like integral of the `sinc` function on *[1, ∞]*, we could issue the following command:

```
In [73]: def f(x): return 1./x
In [74]: quad(f, 1, np.inf weight='sin', wvar=1, limlst=5)
quadpack.py:295: UserWarning: The maximum number of cycles allowed
has been achieved., e.e. of subintervals (a+(k-1)c, a+kc) where c =
(2*int(abs(omega)+1))*pi/abs(omega), for k = 1, 2, ..., lst.  One can
allow more cycles by increasing the value of limlst.  Look at
info['ierlst'] with full_output=1.
  warnings.warn(msg)
Out[74]: (0.636293340511029, 1.3041427865109276)
In [75]: quad(f, 1, np.inf, weight='sin', wvar=1, limlst=50)
Out[75]: (0.6247132564795975, 1.4220476353655983e-08)
```

Numerical multivariate integration

It is also possible to perform multivariate numerical integration on different domains, through application of adaptive Gaussian quadrature rules. In the module `scipy.integrate`, we have to this effect the routines `dblquad` (double integrals), `tplquad` (triple integrals), and `nquad` (integration over multiple variables).

These routines can only compute definite integrals over type I regions:

- In two dimensions, a type I domain can be written in the form $\{(x,y) : a<x<b, f(x)<y<h(x)\}$ for two numbers `a` and `b` and two univariate functions $f(x)$ and $h(x)$.

- In three dimensions, a type I region can be written in the form $\{(x,y,z) : a<x<b, f(x)<y<h(x), q(x,y)<z<r(x,y)\}$ for numbers `a`, `b`, univariate functions $f(x)$, $h(x)$, and bivariate functions $q(x,y)$, $r(x,y)$.

- In more than three dimensions, type I regions can be written sequentially in a similar manner as its double and triple counterparts. The first variable is bounded by two numbers. The second variable is bounded by two univariate functions of the first variable. The third variable is bounded by two bivariate functions of the two first variables, and so on.

Let's run a numerical integration over the function of the example in line In [18]. Note the order in which the different variables must be introduced in the definition of the function to be integrated:

```
In [76]: def f(x, y): return np.exp(-x**2 - y**2)
In [77]: from scipy.integrate import dblquad
In [78]: dblquad(f, 0, np.inf, lambda x:0, lambda x:np.inf)
Out[78]: (0.785398163397,  6.29467149642e-09)
```

Summary

In this chapter, we have mastered all the different methods to compute differentiation and integration of functions. We learned that the scipy libraries have very robust routines to compute approximations of all these operations numerically (wrapping efficient Fortran libraries when necessary). We also learned that it is possible to access other libraries in the SciPy stack to perform the operations in a symbolic or an automatic way.

In the next chapter, we will explore the theory and methodology to solve equations or systems of equations, in the context of nonlinear functions, as well as computing extrema for optimization purposes.

4

Nonlinear Equations and Optimization

In this chapter, we will review two basic operations that are fundamental to the development of Numerical Mathematics: the search of zeros and extrema of real-valued functions.

Motivation

Let's revisit Runge's example from *Chapter 2, Interpolation and Approximation*, where we computed a Lagrange interpolation of Runge's function using eleven equally spaced nodes in the interval from -5 to 5:

```
In [1]: import numpy as np, matplotlib.pyplot as plt; \
   ...: from scipy.interpolate import BarycentricInterpolator
In [2]: def f(t): return 1. / (1. + t**2)
In [3]: nodes = np.linspace(-5, 5, 11); \
   ...: domain = np.linspace(-5, 5, 128); \
   ...: interpolant = BarycentricInterpolator(nodes, f(nodes))
In [4]: plt.figure(); \
   ...: plt.subplot(121); \
   ...: plt.plot(domain, f(domain), 'r-', label='original'); \
   ...: plt.plot(nodes, f(nodes), 'ro', label='nodes'); \
   ...: plt.plot(domain, interpolant1(domain), 'b--',
   ...:          label='interpolant'); \
   ...: plt.legend(loc=9); \
   ...: plt.subplot(122); \
```

```
...: plt.plot(domain, np.abs(f(domain)-interpolant1(domain))); \
...: plt.title('error or interpolation'); \
...: plt.show()
```

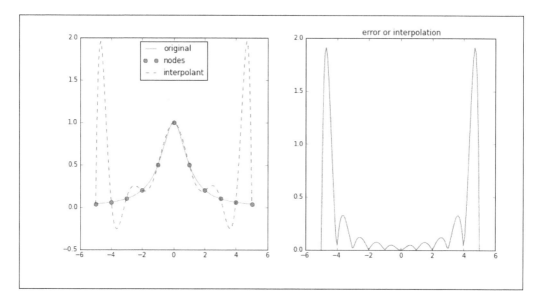

One way to measure the success or failure of this scheme is by computing the uniform norm of the difference between the original function and the interpolation. In this particular case, that norm is close to 2.0. We could approximate this value by performing the following computation on a large set of points in the domain:

```
In [5]: error1a = np.abs(f(domain)-interpolant(domain)).max(); \
    ...: print error1a
```

```
1.91232007608
```

However, this is a crude approximation of the actual error. To compute the true norm, we need a mechanism that calculates the actual maximum value of a function over a finite interval, rather than over a discrete set of points. To perform this operation for the current example, we will use the routine `minimize_scalar` from the module `scipy.optimize`.

Let's solve this problem in two different ways, to illustrate one possible pitfall of optimization algorithms:

- In the first case, we will exploit the symmetry of the problem (both `f` and `interpolator` are even functions) and extract the maximum value of the norm of their difference in the interval from 0 to 5

- In the second case, we perform the same operation over the full interval from -5 to 5.

We will draw conclusions after the computations:

```
In [6]: from scipy.optimize import minimize_scalar
In [7]: def uniform_norm(func, a, b):
   ...:     g = lambda t: -np.abs(func(t))
   ...:     output = minimize_scalar(g, method="bounded",
   ...:                                       bounds=(a, b))
   ...:     return -output.fun
   ...:
In [8]: def difference(t): return f(t) - interpolant(t)
In [9]: error1b = uniform_norm(difference, 0., 5.)
   ...: print error1b
1.9156589182259303
In [10]: error1c = uniform_norm(difference, -5., 5.); \
    ....: print error1c
0.32761452331581842
```

What did just happen? The routine `minimize_scalar` uses an iterative algorithm that got confused by the symmetry of the problem and converged to one local maximum, rather than the requested global maximum.

This first example illustrates one of the topics of this chapter (and its dangers): the computation of constrained extrema for real-valued functions.

The approximation is obviously not very good. A theorem by Chebyshev states that the best polynomial approximation is achieved with a smart choice of nodes—the zeros of the Chebyshev polynomials precisely! We can gather all these roots by using the routine `t_roots` from the module `scipy.special`. In our running example, the best choice of 11 nodes will be based upon the roots of the 11-degree Chebyshev polynomial, properly translated over the interval of the interpolation:

```
In [11]: from scipy.special import t_roots
In [12]: nodes = 5 * t_roots(11)[0]; \
    ....: print nodes
[ -4.94910721e+00   -4.54815998e+00   -3.77874787e+00   -2.70320409e+00
   -1.40866278e+00   -1.34623782e-15    1.40866278e+00    2.70320409e+00
    3.77874787e+00    4.54815998e+00    4.94910721e+00]
In [13]: interpolant = BarycentricInterpolator(nodes, f(nodes))
```

```
In [14]: plt.figure(); \
   ....: plt.subplot(121); \
   ....: plt.plot(domain, f(domain), 'r-', label='original'); \
   ....: plt.plot(nodes, f(nodes), 'ro', label='nodes'); \
   ....: plt.plot(domain, interpolant(domain), 'b--',
   ....:          label='interpolant')); \
   ....: plt.subplot(122); \
   ....: plt.plot(domain, np.abs(f(domain)-interpolant(domain))); \
   ....: plt.title('error or interpolation'); \
   ....: plt.show()
```

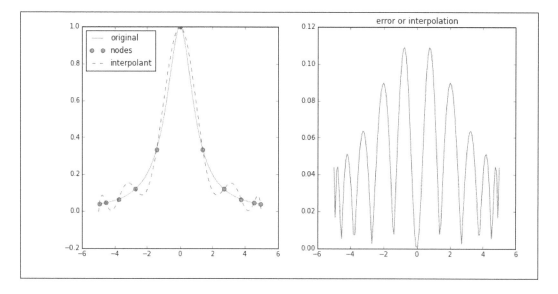

This is a significant improvement in the quality of the interpolator. All thanks to the well-placed nodes that we computed as the roots of a polynomial. Let's compute the uniform norm of this interpolation:

```
In [15]: def difference(t): return f(t) - interpolant(t)
In [16]: error2 = uniform_norm(difference, 0., 2.)
   ....: print error2
0.10915351095
```

For some recurrent cases, such as the example of the zeros of Chebyshev polynomials, the module `scipy.special` has routines that collect those values with prescribed accuracy. For a complete list of those special cases, refer to the online documentation of `scipy.special` at http://docs.scipy.org/doc/scipy/reference/special.html.

For general cases, we would like to have a good set of techniques to get roots. This is precisely the other topic discussed in this chapter.

Non-linear equations and systems

In the solution of linear equations and systems, $f(x) = 0$, we had the choice of using either direct methods or iterative processes. A direct method in that setting was simply the application of an exact formula involving only the four basic operations: addition, subtraction, multiplication, and division. The issues with this method arise when cancellation occurs, mainly whenever sums and subtractions are present. Iterative methods, rather than computing a solution in a finite number of operations, calculate closer and closer approximations to the said solution, improving the accuracy with each step.

In the case of nonlinear equations, direct methods are seldom a good idea. Even when a formula is available, the presence of nonbasic operations leads to uncomfortable rounding errors. Let's see this using a very basic example.

Consider the quadratic equation $ax^2 + bx + c = 0$, with $a = 10^{-10}$, $b = -(10^{10} + 1)/10^{10}$, and $c = 1$. These are the coefficients of the expanded version of the polynomial $p(x) = 10^{-10}(x-1)(x-10^{10})$, with the obvious roots $x = 1$ and $x = 10^{10}$. Notice the behavior of the quadratic formula in the following command:

```
In [1]: import numpy as np
In [2]: a, b, c = 1.0e-10, -(1.0e10 + 1.)/1.0e10, 1.
In [3]: (-b - np.sqrt(b**2 - 4*a*c))/(2*a)
Out[3]: 1.00000000082740371
```

A notable rounding error due to cancellation has spread. It is possible to fix the situation, in this case, by multiplying the numerator and denominator of this formula by the conjugate of its denominator, and using the resulting formula instead:

```
In [4]: 2*c / (-b + np.sqrt(b**2 - 4*a*c))
Out[4]: 1.0
```

Even the algebraic solvers coded in the `sympy` libraries share this defect, as the following example shows:

 The `sympy` libraries have a set of algebraic solvers, and all of them are accessed from the common routine `solve`. Currently, this method solves univariate polynomials, transcendental equations, and a piecewise combination of them. It also solves systems of linear and polynomial equations.

For more information, refer to the official documentation for `sympy` at `http://docs.sympy.org/dev/modules/solvers/solvers.html`.

```
In [5]: from sympy import symbols, solve
In [6]: x = symbols('x', real=True)
In [7]: solve(a*x**2 + b*x + c)
Out[7]: [1.00000000000000, 9999999999.00000]
```

To avoid having to second-guess the accuracy of our solutions or fine-tune each possible formula that solves a nonlinear equation, we can always adopt iterative processes to achieve arbitrarily close approximations.

Iterative methods for univariate functions

Iterative processes for scalar functions can be divided in to three categories:

- **Bracketing methods**, where the algorithms track the endpoints of an interval containing a root. We have the following algorithms:
    - ○ Bisection Method
    - ○ Regula falsi (false position method)

- **Secant methods**, with the following algorithms:
    - ○ The secant method
    - ○ The Newton-Raphson method
    - ○ The interpolation method
    - ○ The inverse interpolation method
    - ○ The fixed-point iteration method

- **Brent method**, which is a combination of the bisection, secant, and inverse interpolation methods.

Now, let's explore the methods included in the SciPy stack.

Bracketing methods

The most basic algorithm is the method of bisection — given a continuous function $f(x)$ in the interval $[a, b]$ satisfying $f(a)f(b) < 0$. This method constructs a sequence of approximations by bisecting intervals and keeping the subinterval where the solution is present. It is a slow process (linear convergence), but it never fails to converge to a solution. In the module `scipy.optimize`, we have one implementation, the routine `bisect`.

Let's explore this method first with our running example. Since the signs of $p(0)$ and $p(2)$ are different, there must be a root in the interval $[0, 2]$:

```
In [8]: from scipy.optimize import bisect
In [9]: p = np.poly1d([a,b,c])
In [10]: bisect(p, 0, 2)
Out[10]: 1.0
```

> Note that we chose to represent $p(x)$ with a `numpy.poly1d` class. Whenever we need to work with polynomials, the optimal way to handle them in SciPy is by means of this class. This ensures evaluation of the polynomials using a Horner scheme, which provides faster computations than with any other lambda representation.
>
> For polynomials with a very high degree, however, the Horner scheme might be inaccurate due to rounding errors from cancellation. Caution must be used in those cases.

One issue with the method of bisection is that it is very sensitive to the choice of initial endpoints, but in general, the quality of the computed solutions can be improved by requesting proper tolerances, as shown in the following example:

```
In [11]: bisect(p, -1, 2)
Out[11]: 1.0000000000002274
In [12]: bisect(p, -1, 2, xtol=1e-15)
Out[12]: 0.9999999999999996
In [13]: bisect(p, -1, 2, xtol=1e-15, rtol=1e-15)
Out[13]: 1.0000000000000009
```

More advanced sets of techniques are based upon regula falsi. Given an interval $[a, b]$ that contains a root of the function $f(x)$, compute the line that goes through the points $(a, f(a))$ and $(b, f(b))$. This line intersects the x axis inside $[a, b]$. We use this point for the next bracketing step. In the module `scipy.optimize`, we have the routine `ridder` (an improvement of regula falsi based on an algorithm developed by C. Ridders), which presents quadratic convergence.

To illustrate the difference in behavior between any solvers, we might use the optional output `RootResult` of each algorithm, as the following session shows:

```
In [14]: soln, info = bisect(p, -1, 2, full_output=True)
In [15]: print "Iterations: {0}".format(info.iterations)
Iterations: 42
In [16]: print "Function calls: {0}".format(info.function_calls)
Function calls: 44
In [17]: from scipy.optimize import ridder
In [18]: soln, info = ridder(p, -1, 2, full_output=True)
In [19]: print "Solution: {0}".format(info.root)
Solution: 1.0
In [20]: print "Iterations: {0}".format(info.iterations)
Iterations: 1
In [21]: print "Function calls: {0}".format(info.function_calls)
Function calls: 4
```

Secant methods

The next step of techniques is based on the secant method and its limit cases. The secant method is very similar to regula falsi computationally. Instead of bracketing the root, we start with any two initial guesses x_0, x_1, and compute the intersection x_2 of the line through $(x_0, f(x_0))$ and $(x_1, f(x_1))$. The next step repeats the same operation on the guesses x_1, x_2 to compute a new approximation x_3, and the process is repeated until a satisfactory approximation to the root is attained.

Improvements on this method can be obtained by employing smarter choices than the secant line to search for intersections with the x axis. The Newton-Raphson method uses a first derivative of $f(x)$ to compute a better intersecting line. The Halley method employs both first and second derivatives of $f(x)$ to compute the intersection of an arc of parabola with the x axis.

The secant method has an order of convergence of approximately 1.61803, while Newton-Raphson is quadratic and Halley is cubic.

For scalar functions, all three methods (secant, Newton, Halley) can be accessed with the common routine `newton` in the module `scipy.optimize`. The obligatory parameters for the routine are the function $f(x)$, together with an initial guess x_0.

Let's work on a more complex example involving the equation *sin(x)/x = 0*:

```
In [22]: from scipy.optimize import newton; \
    ....: from sympy import sin as Sin, pi, diff, lambdify
In [23]: def f(t): return np.sin(t)/t
In [24]: f0 = Sin(x)/x
In [25]: f1prime = lambdify(x, diff(f0, x), "numpy"); \
    ....: f2prime = lambdify(x, diff(f0, x, 2), "numpy")
In [26]: solve(f0, x)
Out[26]: [pi]
In [27]: newton(f, 1)                              # pure secant
Out[27]: 3.1415926535897931
In [28]: newton(f, 1, fprime=f1prime)              # Newton-Raphson
Out[28]: 3.1415926535897931
In [29]: newton(f, 1, fprime=f1prime, fprime2=f2prime)    # Halley
Out[29]: 3.1415926535897931
```

An issue with any of these three methods is that convergence is not always guaranteed. The routine `newton` has a mechanism that prevents the algorithm from going over a certain number of steps and, when this happens, it raises a runtime error that informs us of that situation. A classical example of bad behavior in the Newton-Raphson method and the Halley method occurs with the equation $x^{20} = 1$ (which has the obvious roots $x = 1$ and $x = -1$), if our initial guess happens to be $x = 0.5$:

```
In [30]: solve(x**20 - 1, x)
Out[30]:
[-1,
 1,
 -sqrt(-sqrt(5)/8 + 5/8) + I/4 + sqrt(5)*I/4,
 -sqrt(-sqrt(5)/8 + 5/8) - sqrt(5)*I/4 - I/4,
 sqrt(-sqrt(5)/8 + 5/8) + I/4 + sqrt(5)*I/4,
 sqrt(-sqrt(5)/8 + 5/8) - sqrt(5)*I/4 - I/4,
 -sqrt(sqrt(5)/8 + 5/8) - I/4 + sqrt(5)*I/4,
 -sqrt(sqrt(5)/8 + 5/8) - sqrt(5)*I/4 + I/4,
 sqrt(sqrt(5)/8 + 5/8) - I/4 + sqrt(5)*I/4,
 sqrt(sqrt(5)/8 + 5/8) - sqrt(5)*I/4 + I/4]
In [31]: coeffs = np.zeros(21); \
    ....: coeffs[0] = 1; \
```

```
....: coeffs[20] = -1; \
....: p = np.poly1d(coeffs); \
....: p1prime = p.deriv(m=1); \
....: p2prime = p.deriv(m=2)
In [32]: newton(p, 0.5, fprime=p1prime)
RuntimeError: Failed to converge after 50 iterations, value is
2123.26621974
In [33]: newton(p, 0.5, fprime=p1prime, fprime2=p2prime)
RuntimeError: Failed to converge after 50 iterations, value is
2.65963902147
```

There is yet another technique to approximate solutions to nonlinear scalar equations iteratively, via fixed point iterations. This is very convenient when our equations can be written in the form $x = g(x)$, for example, since the solution to the equation will be a fixed point of the function g.

In general, for any given equation $f(x) = 0$, there is always a convenient way to rewrite it as a fixed point problem $x = g(x)$. The standard way would be to write $g(x) = x + f(x)$, of course, but this does not necessarily provide the best setting. There are many other possibilities out there.

To calculate iteration to a fixed point, we have the routine `fixed_point` in the module `scipy.optimize`. This implementation is based in an algorithm by Steffensen, using a smart convergence acceleration by Aitken:

```
In [34]: def g(t): return np.sin(t)/t + t
In [35]: from scipy.optimize import fixed_point
In [36]: fixed_point(g, 1)
Out[36]: 3.1415926535897913
```

Brent method

Developed by Brent, Dekker, and van Wijngaarten, an even more advanced (and faster) algorithm arises when combining the secant and bisection methods with inverse interpolation. In the module `scipy.optimize`, we have two variations of this algorithm: `brentq` (using inverse quadratic interpolation) and `brenth` (using inverse hyperbolic interpolation). They both start as a bracketing method and require, as input, an interval that contains a root of the function $f(x)$.

Let's compare these two variations of the Brent method to the bracketing methods, with the equation $sin(x)/x = 0$:

```
In [37]: soln, info = bisect(f, 1, 5, full_output=True); \
....: list1 = ['bisect', info.root, info.iterations,
```

```
    ....:              info.function_calls]
In [38]: soln, info = ridder(f, 1, 5, full_output=True); \
    ....: list2 = ['ridder', info.root, info.iterations,
    ....:              info.function_calls]
In [39]: from scipy.optimize import brentq, brenth
In [40]: soln, info = brentq(f, 1, 5, full_output=True); \
    ....: list3 = ['brentq', info.root, info.iterations,
    ....:              info.function_calls]
In [41]: soln, info = brenth(f, 1, 5, full_output=True); \
    ....: list4 = ['brenth', info.root, info.iterations,
    ....:              info.function_calls]
In [42]: for item in [list1, list2, list3, list4]:
    ....:     print "{0}: x={1}. Iterations: {2}. Calls: {3}".
format(*item)
    ....:
bisect: x=3.14159265359. Iterations: 42. Calls: 44
ridder: x=3.14159265359. Iterations: 5. Calls: 12
brentq: x=3.14159265359. Iterations: 10. Calls: 11
brenth: x=3.14159265359. Iterations: 10. Calls: 11
```

Systems of nonlinear equations

In this section, we aim to find solutions of systems of scalar or multivariate functions, $F(X) = 0$, where F represents a finite number N of functions, each of them accepting as a variable a vector X of dimension N.

In the case of systems of algebraic or transcendental equations, symbolic manipulation is a possibility. When the dimensions are too large, it is nonetheless very impractical. A few examples to illustrate this point should suffice.

Let's start with a very easy case that can be readily solved by elimination: the intersection of a circle ($x^2 + y^2 = 16$) with a parabola ($x^2 - 2y = 8$):

```
In [1]: import numpy as np; \
   ...: from sympy import symbols, solve
In [2]: x,y = symbols('x y', real=True)
In [3]: solutions = solve([x**2 + y**2 - 16, x**2 - 2*y -8])
In [4]: for item in solutions:
   ...:     print '({0}, {1})'.format(item[x], item[y])
```

```
    ...:
(0, -4)
(0, -4)
(-2*sqrt(3), 2)
(2*sqrt(3), 2)
```

Now, let's present a harder example. One of the equations is fractional and the other is polynomial: $1/x^4 + 6/y^4 = 6, 2y^4 + 12x^4 = 12x^4y^4$:

```
In [5]: solve([1/x**4 + 6/y**4 - 6, 2*y**4 + 12*x**4 - 12*x**4*y**4])
Out[5]: []
```

No solutions? How about $(1, (6/5))^{1/4}$?

```
In [5]: x0, y0 = 1., (6/5.)**(1/4.)
In [6]: np.isclose(1/x0**4 + 6/y0**4, 6)
Out[6]: True
In [7]: np.isclose(2*y0**4 + 12*x0**4, 12*x0**4*y0**4)
Out[7]: True
```

Only iterative methods can guarantee accurate and fast solutions without exhausting our computational resources. Let's explore some techniques in this direction.

Going from one to several variables brings many computational challenges. Some of the techniques that arise in this context are generalizations from the methods explained for scalar functions in the previous section, but there are many other strategies that exploit the richer structures of spaces with large dimensions. As in the case of solutions of linear equations employing iterative methods, the command of all these techniques involves learning about very advanced topics such as operators in Functional Analysis, Spectral Theory, Krylov subspaces, and so on. This is far beyond the scope of our book.

For a complete description and analysis of all methods, optimal choices of initial guesses, or the construction of successful preconditioners (when used), refer instead to the book *Iterative solutions of nonlinear equations in several variables*, by Ortega and Rheinboldt. It was published in 1970 as a Monograph Textbook for Computational Science and Applied Mathematics by Academic Press, and is still among the best available sources for this topic.

For our analysis of systems of nonlinear equations, we will run all our different methods on a particularly challenging example that tries to determine the values of *x = [x[0], ..., x[8]]*, solving the following system of tridiagonal equations:

```
(3-2*x[0])*x[0]                   -2*x[1]                      = -1
        -x(k-1) + (3-2*x[k])*x[k]            -2*x[k+1] = -1,   k=1,...,7
                          -x[7] + (3-2*x[8])*x[8] = -1
```

We can define such systems as both a purely NumPy function or as a SymPy matrix function (this will help us compute its Jacobian in the future):

```
In [8]: def f(x):
   ...:     output = [(3-2*x[0])*x[0] - 2*x[1] + 1]
   ...:     for k in range(1,8):
   ...:         output += [-x[k-1] + (3-2*x[k])*x[k] - 2*x[k+1] + 1]
   ...:     output += [-x[7] + (3-2*x[8])*x[8] + 1]
   ...:     return output
   ...:
In [9]: from sympy import Matrix, var
In [10]: var('x:9'); \
   ....: X = [x0, x1, x2, x3, x4, x5, x6, x7, x8]
In [11]: F  = Matrix(f(X)); \
   ....: F
Out[11]:
Matrix([
[      x0*(-2*x0 + 3) - 2*x1 + 1],
[-x0 + x1*(-2*x1 + 3) - 2*x2 + 1],
[-x1 + x2*(-2*x2 + 3) - 2*x3 + 1],
[-x2 + x3*(-2*x3 + 3) - 2*x4 + 1],
[-x3 + x4*(-2*x4 + 3) - 2*x5 + 1],
[-x4 + x5*(-2*x5 + 3) - 2*x6 + 1],
[-x5 + x6*(-2*x6 + 3) - 2*x7 + 1],
[-x6 + x7*(-2*x7 + 3) - 2*x8 + 1],
[       -x7 + x8*(-2*x8 + 3) + 1]])
```

All available iterative solvers could be called with the common routine `root` in the module `scipy.optimize`. The routine requires, as obligatory parameters, a left-hand side expression of the system $F(x) = 0$ and an initial guess. To access the different methods, we include the parameter `method`, which can be set to any of the following options:

- `linearmixing`: For linear mixing, a very simple iterative inexact-Newton method that uses a scalar approximation to the Jacobian.

- `diagbroyden`: For diagonal Broyden method, another simple iterative inexact-Newton method that uses a diagonal Broyden approximation to the Jacobian.

- `excitingmixing`: For exciting mixing, one more simple inexact-Newton method, that uses a tuned diagonal approximation to the Jacobian.

- `broyden1`: The good Broyden method is a powerful inexact-Newton method using Broyden's first Jacobian approximation.

- `hybr`: Powell's hybrid method, the most versatile and robust solver available in the SciPy stack, although it is not efficient for systems with large dimensions.

- `broyden2`: The bad Broyden method, similar to the good Broyden method, is another inexact-Newton method that uses Broyden's second Jacobian approximation. It is more apt for large-scale systems.

- `krylov`: The Newton-Krylov method is another inexact-Newton method based on Krylov approximations to the inverse of the Jacobian. It is a top pick for systems with large dimensions.

- `anderson`: This is an extended version of the Anderson mixing method. Together with Newton-Krylov and the bad Broyden method, this is the other weapon of choice for dealing with large scale systems of nonlinear equations.

The implementations are very clever. Except in the case of Powell's hybrid method, the rest uses the same code employing different expressions for the (approximations to the) Jacobian of `f(x)`, `Jacf(x)`. To this effect, there is a python class, `Jacobian`, stored in the module `scipy.optimize.nonlin`, with the following class attributes:

- `.solve(v)`: This returns, for a suitable left-hand-side vector v, the expression $Jacf(x)^{(-1)}*v$

- `.update(x, F)`: This updates the object to `x`, with residual `F(x)`, to guarantee evaluation of the Jacobian at the right location on each step

- `.matvec(v)`: This returns, for a suitable vector v, the product `Jacf(x)*v`

- `.rmatvec(v)`: This returns, for a suitable vector v, the product `Jacf(x).H*v`

- `.rsolve(v)`: This returns, for a suitable vector `v`, the product `(Jacf(x).H)^(-1)*v`

- `.matmat(M)`: For a dense matrix `M` with the appropriate dimensions, this returns the matrix product `Jacf(x).H*M`

- `.todense()`: This forms the dense Jacobian matrix, if ever needed

We seldom need to worry about creating objects in this class. The routine `root` accepts as Jacobians any `ndarray`, sparse matrix, `LinearOperator`, or even callables whose output is any of the previous. It transforms them internally to a `Jacobian` class with the method `asjacobian`. This method is also hosted in the submodule `scipy.optimize.nonlin`.

Simple iterative solvers

We have three very simple inexact-Newton solvers in the SciPy stack which, like the secant method for scalar equations, substitute the Jacobian of a multivariate function with a suitable approximation. These are the methods of linear and exciting mixing, and the diagonal Broyden method. They are fast, but not always reliable—use them at your own risk!

To analyze, in depth, the speed and behavior of these solvers, we will use a callback function to store the steps of convergence. First, the method of linear mixing:

```
In [12]: from scipy.optimize import root
In [13]: root(f, np.zeros(9), method='linearmixing')
Out[13]:
  status: 2
 success: False
     fun: array([  9.73976997e+00,  -1.66208587e+02,   7.98809260e+00,
         -1.66555288e+01,   6.09078392e+01,  -5.57689008e+03,
          5.72527250e+06,  -2.61478262e+13,   3.15410157e+06])
       x: array([  2.85055795e+00,  -8.21972867e+00,   2.28548187e+00,
         -1.17938653e+00,   4.52499108e+00,  -4.30522840e+01,
          8.68604963e+02,  -3.61578590e+06,   4.81211473e+02])
 message: 'The maximum number of iterations allowed has been reached.'
     nit: 1000
```

This is not too promising! If we so desire, we can play around with different tolerances, or the maximum number of iterations allowed. Another option that we could change for this algorithm is the method of searching for optimal lines in the approximation of the Jacobian. This helps determine the step size in the direction given by the said approximation. At this point, we only have three choices: `armijo` (the Armijo-Goldstein condition, the default), `wolfe` (using Philip Wolfe's inequalities), or `None`.

All options passed to any method must be done through a Python dictionary, via the parameter `options`:

```
In [14]: lm_options = {}; \
    ....: lm_options['line_search'] = 'wolfe'; \
    ....: lm_options['xtol'] = 1e-5; \
    ....: lm_options['maxiter'] = 2000
In [15]: root(f, np.zeros(9), method='linearmixing',
    ....:         options=lm_options)
OverflowError: (34, 'Result too large')
```

Now, let's try the method of exciting mixing, with the same initial condition:

```
In [16]: root(f, np.zeros(9), method='excitingmixing')
Out[16]:
  status: 2
 success: False
     fun: array([  1.01316841e+03,  -8.25613756e+05,   4.23367202e+02,
        -7.04594503e+02,   5.53687311e+03,  -2.85535494e+07,
         6.34642518e+06,  -3.11754414e+13,   2.87053285e+06])
       x: array([  1.24211360e+01,  -6.41737121e+02,   1.20299207e+01,
        -1.69891515e+01,   3.26672949e+01,  -3.77759320e+03,
         8.82115576e+02,  -3.94812801e+06,   7.34779049e+02])
 message: 'The maximum number of iterations allowed has been reached.'
     nit: 1000
```

A similar (lack of) success! The relevant options to fine-tune this method are `line_search`, the floating-point value `alpha` (to use `-1/alpha` as the initial approximation to the Jacobian), and the floating-point value `alphamax` (so the entries of the diagonal Jacobian are kept in the range `[alpha,alphamax]`).

Let's try the diagonal Broyden method with the same initial condition:

```
In [17]: root(f, np.zeros(9), method='diagbroyden')
Out[17]:
  status: 2
 success: False
     fun: array([-4.42907377, -0.87124314, -2.61646043, 0.59009568,
-1.34073061,
       -2.06266247, -0.32076522, 0.25120731, 0.0731001 ])
       x: array([ 2.09429178, 1.46991649, -0.06730407, 0.96778603,
0.75367344,
        1.2489588 , 1.46803463, 0.08282948, -0.24223748])
 message: 'The maximum number of iterations allowed has been reached.'
     nit: 1000
```

A poor performance from this method too! We could experiment with the options `line_search` and `alpha` to try to improve the convergence, if needed.

The Broyden method

The good Broyden method is another inexact Newton method that uses an actual Jacobian in the first iteration, but for subsequent iterations, it employs successive rank-one updates. Let's see if we have more luck with our running example:

```
In [18]: root(f, np.zeros(9), method='broyden1')
Out[18]:
  status: 2
 success: False
     fun: array([-111.83572901, -938.30236242, -197.71489446,
-626.93927637,
       -737.43130888, -19.87676004, -107.31583876, -92.32200167,
       -252.26714229])
       x: array([ 6.65222472, 22.1441079 , 9.17971608, 17.78778014,
       19.65632798, 3.43502682, -6.03665297, 6.94424738,
11.87312669])
 message: 'The maximum number of iterations allowed has been reached.'
     nit: 1000
```

To fine-tune this algorithm, besides `line_search` and `alpha`, we also have control over the method of enforcing rank constraints on successive iterations. We could plainly restrict the rank to be not higher than a given threshold, with the optional integer `max_rank`. But even better, we could impose a reduction method that depends on other factors. To do so, we employ the option `reduce_method`.

These are the options:

- `restart`: This reduction method drops all matrix columns.
- `simple`: Only the oldest matrix column is dropped.
- `svd`: Together with the optional integer `to_retain`, this reduction method keeps only the most significant SVD components (up to the integer `to_retain`). If the integer `max_rank` was imposed, a good choice for `to_retain` is usually `max_rank - 2`:

```
In [19]: b1_options = {}; \
   ....: b1_options['max_rank'] = 4; \
   ....: b1_options['reduce_method'] = 'svd'; \
   ....: b1_options['to_retain'] = 2
In [20]: root(f, np.zeros(9), method='broyden1', options=b1_options)
Out[20]:
  status: 2
 success: False
     fun: array([ -1.22226719e+00,   -6.72508500e-02,
-6.31642766e-03,
         -2.24588204e-04,   -1.70786962e-05,   -4.55208297e-05,
         -4.81332054e-06,    1.42432661e-05,   -1.64421441e-05])
       x: array([ 0.87691697,   1.65752568,  -0.16593591,
-0.60204322,  -0.68244063,
         -0.688356  ,  -0.66512492,  -0.59589812,  -0.41638642])
 message: 'The maximum number of iterations allowed has been
reached.'
     nit: 1000
```

Powell's hybrid solver

Among the most successful nonlinear system solvers, we have the hybrid algorithm of Powell, for which there are several Fortran routines named `HYBRID**` in the library `MINPACK`. These routines implement several modified versions of the original algorithm.

The `scipy` routine root, when called with `method='hybr'`, acts as a wrapper to both HYBRID and HYBRIDJ. If an expression for the Jacobian is offered via the optional parameter `jac`, then root calls HYBRIDJ, otherwise, it calls HYBRID. Instead of an actual Jacobian, HYBRID uses an approximation to this operator constructed by forward differences at the starting point.

> For a complete description and analysis of Powell's hybrid algorithm from his author, refer to the article *A Hybrid Method for Nonlinear Equations*, published in 1970 in the journal of Numerical Methods for Nonlinear Algebraic Equations.
>
> For implementation details of the Fortran routines HYBRID and HYBRIDJ, refer to Chapter 4 of the MINPACK user guide at `http://www.mcs.anl.gov/~more/ANL8074b.pdf`.

Let's try again with our elusive example:

```
In [21]: solution = root(f, np.zeros(9), method='hybr')

In [22]: print solution.message

The solution converged.

In [23]: print "The root is approximately x = {0}".format(solution.x)

The root is approximately x = [-0.57065451 -0.68162834 -0.70173245
-0.70421294 -0.70136905 -0.69186564

  -0.66579201 -0.5960342  -0.41641206]

In [24]: print "At that point, it is f(x) = {0}".format(solution.fun)

At that point, it is f(x) = [ -5.10793630e-11   1.00466080e-10
-1.17738708e-10   1.36598954e-10

  -1.25279342e-10   1.10176535e-10  -2.81137336e-11  -2.43449705e-11

   3.32504024e-11]
```

It is refreshing to at least obtain a solution, but we can do better. Let's observe the behavior of `method='hybr'` when a precise Jacobian of *f(x)* is offered. In our case, this operator can be readily computed both symbolically and numerically, as follows:

```
In [25]: F.jacobian(X)
Matrix([
[-4*x0 + 3,        -2,         0,         0,         0,         0,         0,         0,         0],
[       -1, -4*x1 + 3,        -2,         0,         0,         0,         0,         0,         0],
[        0,        -1, -4*x2 + 3,        -2,         0,         0,         0,         0,         0],
[        0,         0,        -1, -4*x3 + 3,        -2,         0,         0,         0,         0],
[        0,         0,         0,        -1, -4*x4 + 3,        -2,         0,         0,         0],
[        0,         0,         0,         0,        -1, -4*x5 + 3,        -2,         0,         0],
[        0,         0,         0,         0,         0,        -1, -4*x6 + 3,        -2,         0],
[        0,         0,         0,         0,         0,         0,        -1, -4*x7 + 3,        -2],
[        0,         0,         0,         0,         0,         0,         0,        -1, -4*x8 + 3]])
```

```
In [26]: def Jacf(x):
   ....:       output = -2*np.eye(9, k=1) - np.eye(9, k=-1)
   ....:       np.fill_diagonal(output, 3-4*x)
   ....:       return output
   ....:
In [27]: root(f, np.zeros(9), jac=Jacf, method='hybr')
   status: 1
  success: True
      qtf: array([ -1.77182781e-09,   2.37713260e-09,   2.68847440e-09,
            -2.24539710e-09,   1.34460264e-09,   8.25783813e-10,
            -3.43525370e-09,   2.36025536e-09,   1.16245070e-09])
     nfev: 25
        r: array([-5.19829211,  2.91792319,  0.84419323, -0.48483853,
0.53965529,
            -0.10614628,  0.23741206, -0.03622988,  0.52590331, -4.93470836,
             2.81299775,  0.2137127 , -0.96934776,  1.03732374, -0.71440129,
             0.27461859,  0.5399114 ,  5.38440026, -1.62750656, -0.6939511 ,
             0.3319492 , -0.11487171,  1.11300907, -0.65871043,  5.3675704 ,
            -2.2941419 , -0.85326984,  1.56089518, -0.01734885,  0.12503146,
             5.42400229, -1.8356058 , -0.64571006,  1.61337203, -0.18691851,
             5.25497284, -2.34515389,  0.34665604,  0.47453522,  4.57813558,
            -2.82915356,  0.98463742,  4.64513056, -1.59583822, -3.76195794])
      fun: array([ -5.10791409e-11,   1.00465636e-10,  -1.17738708e-10,
             1.36598732e-10,  -1.25278898e-10,   1.10176535e-10,
            -2.81135115e-11,  -2.43454146e-11,   3.32505135e-11])
        x: array([-0.57065451, -0.68162834, -0.70173245, -0.70421294,
-0.70136905,
            -0.69186564, -0.66579201, -0.5960342 , -0.41641206])
  message: 'The solution converged.'
     fjac: array([[-0.96956077,  0.19053436,  0.06633131, -0.12548354,
0.00592579,
             0.0356269 ,  0.00473293, -0.0435999 ,  0.01657895],
            [-0.16124306, -0.95068272,  0.1340795 , -0.05374361, -0.08570706,
             0.18508814, -0.04624209, -0.05739585,  0.04797319],
            [ 0.08519719,  0.11476118,  0.97782789, -0.0281114 , -0.08494929,
            -0.05753056, -0.02702655,  0.09769926, -0.04280136],
```

```
        [-0.13529817, -0.0388138 ,  0.03067186,  0.97292228, -0.12168962,
         -0.10168782,  0.0762693 ,  0.0095415 ,  0.04015656],
        [-0.03172212, -0.09996098,  0.07982495,  0.10429531,  0.96154001,
         -0.12901939, -0.13390792,  0.10972049,  0.02401791],
        [ 0.05544828,  0.17833604,  0.03912402,  0.1374237 ,  0.09225721,
          0.93276861, -0.23865212, -0.00446867,  0.09571999],
        [ 0.03507942,  0.00518419,  0.07516435, -0.0317367 ,  0.17368453,
          0.20035625,  0.9245396 , -0.20296261,  0.16065313],
        [-0.05145929, -0.0488773 , -0.08274238, -0.02933344, -0.06240777,
          0.09193555,  0.21912852,  0.96156966, -0.04770545],
        [-0.02071235, -0.03178967, -0.01166247,  0.04865223,  0.05884561,
          0.12459889,  0.11668282, -0.08544005, -0.97783168]])
    njev: 2
```

Observe the clear improvement—we have arrived at the same root, but only in 25 iterations. It was necessary to evaluate the Jacobian only twice.

Large-scale solvers

For large scale systems, the hybrid method is very inefficient, since the strength of the method relies on the internal computation of the inverse of a dense Jacobian matrix. In this setting, we prefer to use more robust inexact-Newton methods.

One of these is the bad Broyden method (broyden2). Anderson mixing (anderson) is also a reliable possibility. However, the most successful is, without a doubt, the Newton-Krylov method (krylov):

```
In [28]: root(f, np.zeros(9), method='krylov')
Out[28]:
   status: 1
  success: True
      fun: array([ -8.48621595e-09,  -1.28607913e-08,  -9.39627043e-10,
         -6.71023681e-10,  -1.12563803e-09,  -3.46839557e-09,
         -7.64257968e-09,  -1.29112268e-08,  -1.26301001e-08])
        x: array([-0.57065452, -0.68162834, -0.70173245, -0.70421294,
-0.70136905,
         -0.69186565, -0.66579202, -0.5960342 , -0.41641207])
  message: 'A solution was found at the specified tolerance.'
      nit: 29
```

We have accomplished a good approximation in only 29 iterations. Improvements are possible through a series of optional parameters. The two crucial options are:

- The iterative solvers for linear equations from the module `scipy.sparse.linalg` used to compute the Krylov approximation to the Jacobian

- A preconditioner for the inner Krylov iteration: a functional expression that approximates the inverse of the Jacobian

> To illustrate the employment of preconditioners, there is a great example in the official documents of SciPy at `http://docs.scipy.org/doc/scipy/reference/tutorial/optimize.html`.
>
> Further explaining the usage of these two options would require a textbook on its own! For the theory behind this technique, refer to the article Jacobian-free Newton-Krylov methods, published by D.A. Knoll and D.E. Keyes in the Journal of Computational Physics. 193, 357 (2003).

Optimization

The optimization problem is best described as the search for a local maximum or minimum value of a scalar-valued function $f(x)$. This search can be performed for all possible input values in the domain of f (and in this case, we refer to this problem as an unconstrained optimization), or for a specific subset of it that is expressible by a finite set of identities and inequalities (and we refer to this other problem as a constrained optimization). In this section, we are going to explore both modalities in several settings.

Unconstrained optimization for univariate functions

We focus on the search for the local minima of a function $f(x)$ in an interval $[a, b]$ (the search for local maxima can then be regarded as the search of the local minima of the function $-f(x)$ in the same interval). For this task, we have the routine `minimize_scalar` in the module `scipy.optimize`. It accepts as obligatory input a univariate function $f(x)$, together with a search method.

Most search methods are based on the idea of bracketing that we used for root finding, although the concept of bracket is a bit different in this setting. In this case, a good bracket is a triple $x < y < z$ where $f(y)$ is less than both $f(x)$ and $f(z)$. If the function is continuous, its graph presents a U-shape on a bracket. This guarantees the existence of a minimum inside of the subinterval $[x, z]$. A successful bracketing method will look, on each successive step, for the target extremum in either $[x, y]$ or $[y, z]$.

Let's construct a simple bracketing method for testing purposes. Assume we have an initial bracket $a < c < b$. By quadratic interpolation, we construct a parabola through the points $(a, f(a))$, $(c, f(c))$, and $(b, f(b))$. Because of the U-shape condition, there must be a minimum (easily computable) for the interpolating parabola, say $(d, f(d))$. It is not hard to prove that the value d lies between the midpoints of the subintervals $[a, c]$, and $[c, b]$. We will use this point d for our next bracketing step. For example, if it happens that $c < d$, then the next bracket will be either $c < d < b$, or $a < c < d$. Easy enough! Let's implement this method:

```
In [1]: import numpy as np; \
   ...: from scipy.interpolate import lagrange; \
   ...: from scipy.optimize import OptimizeResult, minimize_scalar
In [2]: def good_bracket(func, bracket):
   ...:     a, c, b = bracket
   ...:     return (func(a) > func(c)) and (func(b) > func(c))
   ...:
In [3]: def parabolic_step(f, args, bracket, **options):
   ...:     stop = False
   ...:     funcalls = 0
   ...:     niter = 0
   ...:     while not stop:
   ...:         niter += 1
   ...:         interpolator = lagrange(np.array(bracket),
   ...:                                 f(np.array(bracket)))
   ...:         funcalls += 3
   ...:         a, b, c = interpolator.coeffs
   ...:         d = -0.5*b/a
   ...:         if np.allclose(bracket[1], d):
   ...:             minima = d
   ...:             stop = True
   ...:         elif bracket[1] < d:
```

```
    ...:                     newbracket = [bracket[1], d, bracket[2]]
    ...:                     if good_bracket(f, newbracket):
    ...:                         bracket = newbracket
    ...:                     else:
    ...:                         bracket = [bracket[0], bracket[1], d]
    ...:                 else:
    ...:                     newbracket = [d, bracket[1], bracket[2]]
    ...:                     if good_bracket(f, newbracket):
    ...:                         bracket = newbracket
    ...:                     else:
    ...:                         bracket = [bracket[0], d, bracket[1]]
    ...:         return OptimizeResult(fun=f(minima), x=minima,
    ...:                               nit=niter, nfev=funcalls)
```

The output of any minimizing method must be an OptimizeResult object, with at least the attribute x (the solution to the optimization problem). In the example we have just run, the attributes coded in this method are x, fun (the evaluation of *f* at that solution), nit (number of iterations), and nfev (number of functions evaluations needed).

Let's run this method over a few examples:

```
In [4]: f = np.vectorize(lambda x: max(1-x, 2+x))
In [5]: def g(x): return -np.exp(-x)*np.sin(x)
In [6]: good_bracket(f, [-1, -0.5, 1])
Out[6]: True
In [7]: minimize_scalar(f, bracket=[-1, -0.5, 1],
    ...:                 method=parabolic_step)
Out[7]:
   fun: array(1.5000021457670878)
 nfev: 33
  nit: 11
    x: -0.50000214576708779
In [8]: good_bracket(g, [0, 1.2, 1.5])
Out[8]: True
In [9]: minimize_scalar(g, bracket=[0,1.2,1.5],
    ...:                 method=parabolic_step)
```

```
Out[9]:
  fun: -0.32239694192707441
 nfev: 54
  nit: 18
    x: 0.78540558550495643
```

There are two methods already coded for univariate scalar minimization, `golden`, using a golden section search, and `brent`, following an algorithm by Brent and Dekker:

```
In [10]: minimize_scalar(f, method='brent', bracket=[-1, -0.5, 1])
Out[10]:
    fun: array(1.5)
 nfev: 22
  nit: 21
    x: -0.5
In [11]: minimize_scalar(f, method='golden', bracket=[-1, -0.5, 1])
Out[11]:
 fun: array(1.5)
    x: -0.5
 nfev: 44
In [12]: minimize_scalar(g, method='brent', bracket=[0, 1.2, 1.5])
Out[12]:
  fun: -0.32239694194483443
 nfev: 11
  nit: 10
    x: 0.78539817180087257
In [13]: minimize_scalar(g, method='golden', bracket=[0, 1.2, 1.5])
Out[13]:
  fun: -0.32239694194483448
    x: 0.785398157328422 6
 nfev: 43
```

Constrained optimization for univariate functions

Although the bracket included in the routine `minimize_scalar` already places a constraint on the function, it is feasible to force the search for a true minimum inside of a suitable interval for which no bracket can be easily found:

```
In [14]: minimize_scalar(g, method='bounded', bounds=(0, 1.5))
Out[14]:
  status: 0
    nfev: 10
 success: True
     fun: -0.32239694194483415
       x: 0.78539813414299553
 message: 'Solution found.'
```

Unconstrained optimization for multivariate functions

Except in the cases of minimization by brute force or by basin hopping, we can perform all other searches with the common routine `minimize` from the module `scipy.optimize`. The parameter method, as with its univariate counterpart, takes care of selecting the algorithm employed to achieve the extremum. There are several well-known algorithms already coded, but we also have the possibility of implementing our own ideas via a suitable custom-made method.

In this section, we will focus on the description and usage of the coded implementations. The same technique that we employed in the construction of custom methods for `minimize_scalar` is valid here, with the obvious challenges that the extra dimensions bring.

To compare all the different methods, we are going to run them against a particularly challenging function: Rocksenbrock's parabolic valley (also informally referred to as the banana function). The module `scipy.optimize` has NumPy versions of this function, as well as its Jacobian and Hessian:

```
In [15]: from scipy.optimize import rosen; \
   ....: from sympy import var, Matrix, solve, pprint
In [16]: var('x y')
Out[16]: (x, y)
In [17]: F = Matrix([rosen([x, y])]); \
```

```
    ....: pprint(F)
[(-x + 1)² + 100.0(-x² + y)²]
In [18]: X, Y = np.mgrid[-1.25:1.25:100j, -1.25:1.25:100j]
In [19]: def f(x,y): return rosen([x, y])
In [20]: import matplotlib.pyplot as plt, matplotlib.cm as cm; \
    ....: from mpl_toolkits.mplot3d.axes3d import Axes3D
In [21]: plt.figure(); \
    ....: plt.subplot(121, aspect='equal'); \
    ....: plt.contourf(X, Y, f(X,Y), levels=np.linspace(0,800,16),
    ....:              cmap=cm.Greys)
    ....: plt.colorbar(orientation='horizontal')
    ....: plt.title('Contour plot')
    ....: ax = plt.subplot(122, projection='3d', aspect='equal')
    ....: ax.plot_surface(X, Y, f(X,Y), cmap=cm.Greys, alpha=0.75)
    ....: plt.colorbar(orientation='horizontal')
    ....: plt.title('Surface plot')
    ....: plt.show()
```

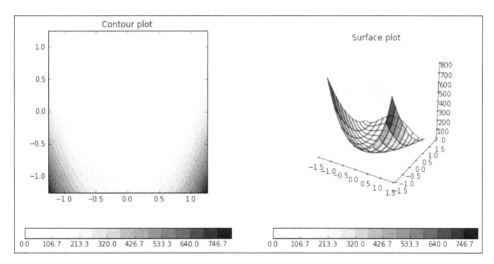

The figure shows a large area (in the shape of a banana) that might contain local minima. Techniques of multivariate calculus help us locate all of the critical points precisely, instead of relying on intuition. We need first to compute the Jacobian and Hessian of the function:

```
In [22]: JacF = F.jacobian([x, y]); \
    ....: pprint(JacF)
[- 400.0·x·(- x² + y) + 2·x - 2 - 200.0·x²  + 200.0·y]
In [23]: HesF = JacF.jacobian([x, y]); \
    ....: pprint(HesF)
[1200.0·x² - 400.0·y + 2 -400.0·x
        -400.0·x              200.0  ]
In [24]: solve(JacF)
Out[24]: [{x: 1.00000000000000, y: 1.00000000000000}]
In [25]: HesF.subs({x: 1.0, y: 1.0})
Out[25]:
Matrix([
[ 802.0, -400.0],
[-400.0,  200.0]])
In [26]: _.det()
Out[26]: 400.000000000000
```

These computations show that there is only one critical point at $(1, 1)$. Unequivocally, this point presents a local minimum at that location.

Trying to compute the critical points with this technique for a Rosenbrock function in higher dimensions, while doable, is computationally intense. Moving to four dimensions, for example, takes a decent computer about half a minute:

```
In [27]: var('x:4'); \
    ....: X = [x0, x1, x2, x3]; \
    ....: F = Matrix([rosen(X)])
In [28]: %time solve(F.jacobian(X))
CPU times: user 36.6 s, sys: 171 ms, total: 36.8 s
Wall time: 36.7 s
Out[28]:
[{x₀:1.0,x₁:1.0,x₂:1.0,x₃:1.0}]
```

For large dimensions, the search for global minima can be done by brute-force; not very elegant, but it gets the job done. A brute-force algorithm is able to track global minima (or approximate it to satisfactory precision). We can call this method with the routine `brute` in the module `scipy.optimize`. The obligatory parameters are the function to be minimized, together with a description of the domain where we will apply the optimization. This domain is best coded as a tuple of slices. For example, to search for a global minimum of the Rosenbrock function in four variables, where each variable is bounded in absolute value by three, we could issue this command:

```
In [29]: from scipy.optimize import brute
In [30]: interval = slice(-3, 3, 0.25); \
   ....: box = [interval] * 4
In [31]: %time brute(rosen, box)
CPU times: user 13.7 s, sys: 6 ms, total: 13.7 s
Wall time: 13.7 s
Out[31]: array([ 1.,   1.,   1.,   1.])
```

Still quite a slow process! To achieve speed, it is always better to use iterative methods. The search for minima in this setting is achieved according to several schema (and combinations of these):

- **The stochastic methods**: These are methods suitable for the search of actual global minima. They generate and use random variables. In the module `scipy.optimize`, we have two exponents of this category:

  - One is the basin-hopping algorithm, called with the routine `basinhopping` in the module `scipy.optimize`. The implementation has an acceptance test given by the Metropolis criterion of a standard **Monte-Carlo simulation**.

  - Another is a deprecated version of the method of simulated annealing, called with `method='Anneal'`. This is a variation of a Monte-Carlo simulation. It is useful for optimization problems where the search space is discrete and large.

- **Deterministic algorithms that exclusively employ function evaluations**: These are basically performed by successive linear minimizations in different directions. In the module `scipy.optimize`, we have two methods complying with this philosophy:
    - Powell's method based on the unidimensional Brent's minimization. We call it with `method='Powell'`.
    - The downhill simplex algorithm, also known as the amoeba method, created by Nelder and Mead in 1965. We call it with `method='Nelder-Mead'`.

- **The Newton methods**: These are deterministic algorithms on differentiable functions that mimic multivariate calculus to search for critical points. In a nutshell, we seek for at least one critical point whose Hessians satisfy the conditions for the local minimum. These algorithms employ both Jacobian and Hessian evaluations. Because of the complexity of these expressions in general, approximations to both operators are usually implemented instead. When this is the case, we refer to these methods as quasi-Newton methods. In the module `scipy.optimize`, we have the quasi-Newton method of **Broyden, Fletcher, Goldfarb, and Shanno (BFGS)**, which uses exclusively first derivatives. We call it with `method='BFGS'`.

- **The conjugate gradient methods**: Here, we have three variants:
    - A variant of the Fetcher-Reeves algorithm is to implement a pure conjugate gradient, written by Polak and Ribiere. It uses exclusively first derivatives and is called with the `method ='CG'`.
    - A combination of the conjugate gradient with a Newton method, the truncated Newton method, which we call with `method='Newton-CG'`.
    - Two different versions of the Newton conjugate gradient trust-region algorithm, which use the idea of trust-regions to more effectively bound the location of the possible minima. We call them with `method='dogleg'` and `method='trust-ncg'`.

Let's browse through these methods.

The stochastic methods

Let's find the global minimum of the Rosenbrock function of nine variables using the technique of basin hopping:

```
In [32]: from scipy.optimize import minimize, basinhopping
In [33]: %time basinhopping(rosen, np.zeros(9))
```

```
CPU times: user 4.59 s, sys: 7.17 ms, total: 4.6 s
Wall time: 4.6 s
Out[33]:
                    nfev: 75633
 minimization_failures: 52
                     fun: 2.5483642615054407e-11
                       x: array([ 0.99999992,  0.99999994,  0.99999992,
0.99999981,  0.99999962,
          0.99999928,  0.99999865,  0.9999972 ,  0.99999405])
                 message: ['requested number of basinhopping iterations
completed successfully']
                   njev: 6820
                    nit: 100
```

Let's compare to the behavior of (the deprecated) simulated annealing:

```
In [34]: minimize(rosen, np.zeros(9), method='Anneal')
Out[34]:
  status: 5
 success: False
  accept: 19
    nfev: 651
       T: 1130372817.0369582
     fun: 707171392.44894326
       x: array([ 11.63666756, -24.41186725,  48.26727994,   3.97730959,
       -31.52658563,  18.00560694,   1.22589971,  21.97577333,
-43.9967434 ])
 message: 'Final point not the minimum amongst encountered points'
     nit: 12
```

Deterministic algorithms that exclusively employ function evaluations

Let's compare the results of the Powell method with the downhill simplex method:

```
In [35]: minimize(rosen, np.zeros(9), method='Powell')
Out[35]:
  status: 0
 success: True
```

```
   direc: array([[ -9.72738085e-06,    2.08442100e-05,    2.06470355e-05,
          4.39487337e-05,    1.29109966e-04,    1.98333214e-04,
          3.66992711e-04,    7.00645878e-04,    1.38618490e-03],
        [ -6.95913466e-06,   -7.25642357e-07,   -2.39771165e-06,
          4.10148947e-06,   -6.17293950e-06,   -6.53887928e-06,
          -1.06472130e-05,   -5.23030557e-06,   -2.28609232e-06],
        [  0.00000000e+00,    0.00000000e+00,    1.00000000e+00,
          0.00000000e+00,    0.00000000e+00,    0.00000000e+00,
          0.00000000e+00,    0.00000000e+00,    0.00000000e+00],
        [  1.23259262e-06,    9.30817407e-07,    2.48075497e-07,
          -7.07907998e-07,   -2.01233105e-07,   -1.10513430e-06,
          -2.57164619e-06,   -2.58316828e-06,   -3.89962665e-06],
        [  6.07328675e-02,    8.51817777e-02,    1.30174960e-01,
          1.71511253e-01,    9.72602622e-02,    1.47866889e-02,
          1.12376083e-03,    5.35386263e-04,    2.04473740e-04],
        [  0.00000000e+00,    0.00000000e+00,    0.00000000e+00,
          0.00000000e+00,    0.00000000e+00,    1.00000000e+00,
          0.00000000e+00,    0.00000000e+00,    0.00000000e+00],
        [  3.88222708e-04,    8.26386166e-04,    5.56913200e-04,
          3.08319925e-04,    4.45122275e-04,    2.66513914e-03,
          6.31410713e-03,    1.24763367e-02,    2.45489699e-02],
        [  0.00000000e+00,    0.00000000e+00,    0.00000000e+00,
          0.00000000e+00,    0.00000000e+00,    0.00000000e+00,
          0.00000000e+00,    1.00000000e+00,    0.00000000e+00],
        [ -2.82599868e-13,    2.33068676e-13,   -4.23850631e-13,
          -1.23391999e-12,   -2.41224441e-12,   -5.08909225e-12,
          -9.92053051e-12,   -2.07685498e-11,   -4.10004188e-11]])
    nfev: 6027
     fun: 3.1358222279861171e-21
       x: array([ 1.,   1.,   1.,   1.,   1.,   1.,   1.,   1.,   1.])
 message: 'Optimization terminated successfully.'
     nit: 56
In [36]: minimize(rosen, np.zeros(9), method='Nelder-Mead')
  status: 1
    nfev: 1800
 success: False
     fun: 4.9724099905503065
       x: array([ 0.85460488,   0.70911132,   0.50139591,   0.24591886,
0.06234451,
```

```
            -0.01112426,  0.02048509,  0.03266785, -0.01790827])
  message: 'Maximum number of function evaluations has been exceeded.'
      nit: 1287
```

The Broyden-Fletcher-Goldfarb-Shanno quasi-Newton method

Let's observe the behavior of this algorithm on our running example:

```
In [37]: minimize(rosen, np.zeros(9), method='BFGS')
Out[37]:
   status: 0
  success: True
     njev: 83
     nfev: 913
 hess_inv: array([[ 1.77874730e-03,  1.01281617e-03,  5.05884211e-04,
            3.17367120e-04,  4.42590321e-04,  7.92168518e-04,
            1.52710497e-03,  3.06357905e-03,  6.12991619e-03],
          [ 1.01281617e-03,  1.91057841e-03,  1.01489866e-03,
            7.31748268e-04,  8.86058826e-04,  1.44758106e-03,
            2.76339393e-03,  5.60288875e-03,  1.12489189e-02],
          [ 5.05884211e-04,  1.01489866e-03,  2.01221575e-03,
            1.57845668e-03,  1.87831124e-03,  3.06835450e-03,
            5.86489711e-03,  1.17764144e-02,  2.35964978e-02],
          [ 3.17367120e-04,  7.31748268e-04,  1.57845668e-03,
            3.13024681e-03,  3.69430791e-03,  6.16910056e-03,
            1.17339522e-02,  2.33374859e-02,  4.66640347e-02],
          [ 4.42590321e-04,  8.86058826e-04,  1.87831124e-03,
            3.69430791e-03,  7.37204979e-03,  1.23036988e-02,
            2.33709766e-02,  4.62512165e-02,  9.24474614e-02],
          [ 7.92168518e-04,  1.44758106e-03,  3.06835450e-03,
            6.16910056e-03,  1.23036988e-02,  2.48336778e-02,
            4.71369608e-02,  9.29927375e-02,  1.85683729e-01],
          [ 1.52710497e-03,  2.76339393e-03,  5.86489711e-03,
            1.17339522e-02,  2.33709766e-02,  4.71369608e-02,
            9.44348689e-02,  1.86490477e-01,  3.72360210e-01],
          [ 3.06357905e-03,  5.60288875e-03,  1.17764144e-02,
            2.33374859e-02,  4.62512165e-02,  9.29927375e-02,
            1.86490477e-01,  3.73949424e-01,  7.46959044e-01],
          [ 6.12991619e-03,  1.12489189e-02,  2.35964978e-02,
```

```
        4.66640347e-02,    9.24474614e-02,    1.85683729e-01,
        3.72360210e-01,    7.46959044e-01,    1.49726446e+00]])
    fun: 6.00817150312557e-11
      x: array([ 0.99999993,  0.99999986,  0.99999976,  0.99999955,
0.99999913,
        0.99999832,  0.99999666,  0.99999334,  0.99998667])
message: 'Optimization terminated successfully.'
    jac: array([  5.23788826e-06,  -5.45925187e-06,  -1.35362172e-06,
        8.75480656e-08,  -9.45374358e-06,   7.31889131e-06,
        3.34352248e-07,  -7.24984749e-07,   2.02705630e-08])
```

 Note that this method employs more iterations, but much fewer function evaluations than the method of Powell (including Jacobian evaluations). Accuracy is comparable, but the gain in complexity and speed in remarkable.

The conjugate gradient method

The pure conjugate gradient method works best with functions with a clear, unique critical point, and where the range of the slopes is not too large. Multiple stationary points tend to confuse the iterations, and too steep slopes (larger than 1000) result in terrible rounding errors.

Without offering an expression for the Jacobian, the algorithm computes a decent approximation of this operator to compute the first derivatives:

```
In [38]: minimize(rosen, np.zeros(9), method='CG')
Out[38]:
  status: 0
 success: True
    njev: 326
    nfev: 3586
     fun: 1.5035665428352255e-10
       x: array([ 0.9999999 ,  0.99999981,  0.99999964,  0.99999931,
0.99999865,
        0.99999733,  0.9999947 ,  0.99998941,  0.99997879])
 message: 'Optimization terminated successfully.'
     jac: array([ -1.48359492e-06,   2.95867756e-06,   1.71067556e-06,
        -1.83617409e-07,  -2.47616618e-06,  -5.34951641e-06,
         2.50389338e-06,  -2.37918319e-06,  -3.86667920e-06])
```

Including an actual Jacobian improves matters greatly. Note the improvement in the evaluation of the found minimum (`fun`):

```
In [39]: from scipy.optimize import rosen_der
In [40]: minimize(rosen, np.zeros(9), method='CG', jac=rosen_der)
Out[40]:
   status: 0
  success: True
     njev: 406
     nfev: 406
      fun: 8.486856765134401e-12
        x: array([ 0.99999998,  0.99999996,  0.99999994,  0.99999986,
0.99999969,
           0.99999938,  0.99999875,  0.9999975 ,  0.999995  ])
  message: 'Optimization terminated successfully.'
      jac: array([  1.37934336e-06,  -9.03688875e-06,   8.53289049e-06,
           9.77779178e-06,  -2.63022111e-06,  -1.02087919e-06,
          -6.55712127e-06,  -1.71887373e-06,  -9.12268328e-07])
```

The truncated Newton method requires a precise Jacobian to work:

```
In [41]: minimize(rosen, np.zeros(9), method='Newton-CG')
ValueError: Jacobian is required for Newton-CG method
In [38]: minimize(rosen, np.zeros(9), method='Newton-CG', jac=rosen_der)
Out[41]:
   status: 0
  success: True
     njev: 503
     nfev: 53
      fun: 5.231613200425767e-08
        x: array([ 0.99999873,  0.99999683,  0.99999378,  0.99998772,
0.99997551,
           0.99995067,  0.99990115,  0.99980214,  0.99960333])
  message: 'Optimization terminated successfully.'
     nhev: 0
      jac: array([  6.67155399e-06,   2.50927306e-05,   1.03398234e-04,
           4.09953321e-04,   1.63524314e-03,   6.48667316e-03,
          -1.91779902e-03,  -2.81972861e-04,  -5.67500380e-04])
```

The methods using trust regions require an exact expression for the Hessian:

```
In [42]: from scipy.optimize import rosen_hess
In [43]: minimize(rosen, np.zeros(9), method='dogleg',
   ....:                     jac=rosen_der, hess=rosen_hess)
Out[43]:
  status: 0
 success: True
    njev: 25
    nfev: 29
     fun: 9.559277795967234e-19
       x: array([ 1.,   1.,   1.,   1.,   1.,   1.,   1.,   1.,   1.])
 message: 'Optimization terminated successfully.'
    nhev: 24
     jac: array([  3.84137166e-14,    3.00870439e-13,    1.10489395e-12,
         4.32831548e-12,    1.72455383e-11,    6.77315980e-11,
         2.48459919e-10,    6.62723207e-10,   -1.52775570e-09])
     nit: 28
In [44]: minimize(rosen, np.zeros(9), method='trust-ncg',
   ....:                     jac=rosen_der, hess=rosen_hess)
Out[44]:
  status: 0
 success: True
    njev: 56
    nfev: 67
     fun: 3.8939669818289621e-18
       x: array([ 1.,   1.,   1.,   1.,   1.,   1.,   1.,   1.,   1.])
 message: 'Optimization terminated successfully.'
    nhev: 55
     jac: array([  2.20490293e-13,    5.57109914e-13,    1.77013959e-12,
        -9.03965791e-12,   -3.05174774e-10,    3.03425818e-09,
         1.49134067e-08,    6.32240935e-08,   -3.64210218e-08])
     nit: 66
```

Note the huge improvement in terms of accuracy, iterations, and function evaluations over the previous methods! The obvious drawback is that quite often it is very challenging to obtain good representations of the Jacobian or Hessian operators.

Constrained optimization for multivariate functions

Take, for example, the minimization of the plane function $f(x,y) = 5x - 2y + 4$ over the circle $x^2 + y^2 = 4$. Using SymPy, we can implement the technique of Lagrange multipliers:

```
In [45]: F = Matrix([5*x - 2*y + 4]); \
   ....: G = Matrix([x**2 + y**2 - 4])      # constraint
In [46]: var('z'); \
   ....: solve(F.jacobian([x, y]) - z * G.jacobian([x, y]))
Out[46]: [{x: 5/(2*z), y: -1/z}]
In [47]: soln = _[0]; \
   ....: solve(G.subs(soln))
Out[47]: [{z: -sqrt(29)/4}, {z: sqrt(29)/4}]
In [48]: zees = _; \
   ....: [(soln[x].subs(item), soln[y].subs(item)) for item in zees]
Out[48]:
[(-10*sqrt(29)/29, 4*sqrt(29)/29), (10*sqrt(29)/29, -4*sqrt(29)/29)]
```

Not too bad! On top of this constraint, we can further impose another condition in the form of an inequality. Think of the same problem as before, but constraining to half a circle instead: $y > 0$. This being the case, the new result will be only the point with coordinates $x = -10\sqrt{(29)}/29 = -1.8569533817705186$ and $y = 4\sqrt{(29)}/29 = 0.74278135270820744$.

It is, of course, possible to address this problem numerically. In the module `scipy.optimize`, we have basically three methods, all of which can be called from the common routine `minimize`:

- The large-scale bound-constrained optimization based on the BFGS algorithm (we call it with `method='L-BFGS-B'`). The implementation is actually a wrapper for a FORTRAN routine with the same name, written by Ciyou Zhu, Richard Byrd, and Jorge Nocedal (for details, see for example, R. H. Byrd, P. Lu, and J. Nocedal. *A Limited Memory Algorithm for Bound Constrained Optimization*, (1995), *SIAM Journal on Scientific and Statistical Computing*, 16, 5, pp. 1190-1208).

- A constrained-based algorithm based on the truncated Newton method (we call it with `method='TNC'`). This implementation is similar to the one we called with `method='Newton-CG'`, except this version is a wrapper for a C routine.

- A constrained optimization by linear approximation (called with `method='COBYLA'`). This implementation wraps a FORTRAN routine with the same name.

- A minimization method based on sequential least squares programming (`method='SLSQP'`). The implementation is a wrapper for a FORTRAN routine with the same name, written by Dieter Kraft.

Let's use our running example to illustrate how to input different constraints. We implement these as a dictionary or a tuple of dictionaries—each entry in the tuple represents either an identity (`'eq'`), or an inequality (`'ineq'`), together with a functional expression (in the form of a `ndarray` when necessary) and the corresponding derivative of it:

```
In [49]: def f(x): return 5*x[0] - 2*x[1] + 4
In [50]: def jacf(x): return np.array([5.0, -2.0])
In [51]: circle = {'type': 'eq',
   ....:                 'fun': lambda x: x[0]**2 + x[1]**2 - 4.0,
   ....:                 'jac': lambda x: np.array([2.0 * x[0],
                                                    2.0 * x[1]])}
In [52]: semicircle = ({'type': 'eq',
   ....:                  'fun': lambda x: x[0]**2 + x[1]**2 - 4.0,
   ....:                  'jac': lambda x: np.array([2.0 * x[0],
                                                     2.0 * x[1]])},
   ....:                 {'type': 'ineq',
   ....:                  'fun': lambda x: x[1],
   ....:                  'jac': lambda x: np.array([0.0, 1.0])})
```

The constraints are fed to the routine `minimize` through the parameter `constraints`. The initial guess must satisfy the constraints too, otherwise, the algorithm fails to converge to anything meaningful:

```
In [53]: minimize(f, [2,2], jac=jacf, method='SLSQP', constraints=circle)
Out[53]:
  status: 0
 success: True
    njev: 11
    nfev: 13
     fun: -6.7703296142789693
       x: array([-1.85695338,  0.74278135])
 message: 'Optimization terminated successfully.'
```

```
    jac: array([ 5., -2.,  0.])
    nit: 11
In [54]: minimize(f, [2,2], jac=jacf, method='SLSQP',
constraints=semicircle)
Out[54]:
  status: 0
 success: True
    njev: 11
    nfev: 13
     fun: -6.7703296142789693
       x: array([-1.85695338,  0.74278135])
 message: 'Optimization terminated successfully.'
     jac: array([ 5., -2.,  0.])
     nit: 11
```

Summary

In this chapter, we have mastered two of the most challenging processes in computational mathematics—the search for roots and extrema of functions. You learned about the symbolic and numerical methods to address these problems in several settings, and how to avoid common pitfalls by gathering enough information about the functions.

In the next chapter, we will explore some techniques for solving differential equations.

5

Initial Value Problems for Ordinary Differential Equations

Initial value problems for ordinary differential equations (or systems) hardly need any motivation. They arise naturally in almost all sciences. In this chapter, we will focus on mastering numerical methods to solve these equations.

Throughout the chapter, we will explore all the techniques implemented in the SciPy stack through three common examples:

- The trivial differential equation of the first order $y'(t) = y(t)$ with the initial condition as $y(0) = 1$. The actual solution is $y(t) = e^t$.

- A more complex differential equation of first order: a Bernoulli equation $ty'(t) + 6y(t) = 3ty(t)^{3/4}$, with the initial condition $y(1) = 1$. The actual solution is $y(t) = (3t^{5/2} + 7)^4/(10000t^6)$.

- To illustrate the behavior with autonomous systems (where the derivatives do not depend on the time variable), we use a Lotka-Volterra model for one predator and one prey, $y_0'(t) = y_0(t) - 0.1\ y_0(t)\ y_1(t)$ and $y_1'(t) = -1.5\ y_1(t) + 0.075\ y_0(t)\ y_1(t)$ with the initial condition $y_0(0) = 10$ and $y_1(0) = 5$ (representing 10 prey and 5 predators at the initial time).

 Higher-order differential equations can always be transformed into (non-necessarily autonomous) systems of differential equations. In turn, nonautonomous systems of differential equations can always be turned into autonomous by including a new variable in a smart way. The techniques to accomplish these transformations are straightforward, and explained in any textbook on differential equations.

We have the possibility of solving some differential equations analytically via SymPy. Although this is hardly the best way to solve initial value problems computationally—even when analytical solutions are available—we will illustrate some examples for completion. Reliable solvers are numerical in nature, and in this setting, there are mainly two ways to approach this problem—through analytic approximation methods, or with discrete-variable methods.

Symbolic solution of differential equations

Symbolic treatment of a few types of differential equations is coded in the SciPy stack through the module `sympy.solvers.ode`. At this point, only the following equations are accessible with this method:

- First order separable
- First order homogeneous
- First order exact
- First order linear
- First order Bernoulli
- Second order Liouville
- Any order linear equations with constant coefficients

In addition to these, other equations might be solvable with the following techniques:

- A power series solution for the first or second order equations (the latter only at ordinary and regular singular points)
- The lie group method for the first order equations

Let's see these techniques in action with our one-dimensional examples, $y' = y$ and the Bernoulli equation. Note the method of inputting a differential equation. We write it in the form $F(t,y,y') = 0$, and we feed the expression $F(t,y,y')$ to the solver (see line 3 that follows). Also, notice how we code derivatives of a function using SymPy; the expression `f(t).diff(t)` represents the first derivative of $f(t)$, for instance:

```
In [1]: from sympy.solvers import ode
In [2]: t = symbols('t'); \
   ...: f = Function('f')
In [3]: equation1 = f(t).diff(t) - f(t)
In [4]: ode.classify_ode(equation1)
Out[4]:
('separable',
```

```
'1st_exact',
'1st_linear',
'almost_linear',
'1st_power_series',
'lie_group',
'nth_linear_constant_coeff_homogeneous',
'separable_Integral',
'1st_exact_Integral',
'1st_linear_Integral',
'almost_linear_Integral')
```

 Note that some of the methods have a variant labeled with the suffix `_Integral`. This is a clever mechanism that allows us to complete the solution without actually computing the needed integrals. This is useful when facing expensive or impossible integrals.

The equation has been classified as a member of several types. We can now solve it according to the proper techniques of the corresponding type. For instance, we choose to solve this equation by first assuming that it is separable, and then by computing an approximation of degree four *(n=4)* to the solution with a power series representation *(hint='1st_power_series')* around x0=0:

```
In [5]: ode.dsolve(equation1, hint='separable')
Out[5]: f(t) == C1*exp(t)
In [6]: ode.dsolve(equation1, hint='1st_power_series', n=4, x0=0.0)
Out[6]: f(t) == C0 + C0*t + C0*t**2/2 + C0*t**3/6 + O(t**4)
```

Solving initial value problems is also possible, but only for solutions computed as a power series of first order differential equations:

```
In [7]: ode.dsolve(equation1, hint='1st_power_series', n=3, x0=0,
   ...:                   ics={f(0.0): 1.0})
Out[7]: f(t) == 1.0 + 1.0*t + 0.5*t**2 + O(t**3)
```

Let's explore the second example with these techniques:

```
In [8]: equation2 = t*f(t).diff(t) + 6*f(t) - 3*t*f(t)**(0.75)
In [9]: ode.classify_ode(equation2)
Out[9]: ('Bernoulli', 'lie_group', 'Bernoulli_Integral')
In [10]: dsolve(equation2, hint='Bernoulli')
```

```
Out[10]: f(t) == (t**(-1.5)*(C1 + 0.3*t**2.5))**4.0
In [11]: dsolve(equation2, hint=lie_group')
Out[11]: f(t) == -5/(3*t*(C1*t**5 - 1))
[f(t) == 6.25e-6*(t**6*(625.0*C1**4 + 5400.0*C1*t**5 + 1296.0*t**10)
 - 120.0*sqrt(C1*t**17*(25.0*C1 + 36.0*t**5)**2))/t**12,
 f(t) == 6.25e-6*(t**6*(625.0*C1**4 + 5400.0*C1*t**5 + 1296.0*t**10)
 + 120.0*sqrt(C1*t**17*(25.0*C1 + 36.0*t**5)**2))/t**12]
```

Of course, although the functional expressions of both solutions in lines 10 and 11 are different, they represent the same function.

> For more information on how to use these techniques and code your own symbolic solvers, refer to the excellent documentation from the official SymPy pages at http://docs.sympy.org/dev/modules/solvers/ode.html.

Analytic approximation methods

Analytic approximation methods try to compute approximations to the exact solutions on suitable domains, in the form of truncated series expansions over a system of basis functions. In the SciPy stack, we have an implementation based on the Taylor series, through the routine odefun in the module sympy.mpmath.

> mpmath is a Python library for arbitrary-precision floating-point arithmetic, hosted inside the sympy module. Although it is independent of the numpy machinery, they both work well together.
>
> For more information about this library, read the official documentation at http://mpmath.org/doc/current/.

Let's see it in action, first with our trivial example $y'(t) = y(t)$, $y(0) = 1$. The key here is to assess the speed and the accuracy of the approximation, as compared to the actual solution in the interval $[0, 1]$. Its syntax is very simple, we assume the equation is always in the form of $y' = F$, and we provide the routine odefun with this functional F and the initial conditions (in this case, 0 for the t-value, and 1 for the y-value):

```
In [1]: import numpy as np, matplotlib.pyplot as plt; \
   ...: from sympy import mpmath
In [2]: def F(t, y): return y
In [3]: f = odefun(F, 0, 1)
```

We proceed to compare the results of the solver `f` with the actual analytical solution `np.exp`:

```
In [4]: t = np.linspace(0, 1, 1024); \
   ...: Y1 = np.vectorize(f)(t); \
   ...: Y2 = np.exp(t)
In [5]: (np.abs(Y1-Y2)).max()
Out[5]: mpf('2.2204460492503131e-16')
```

Let's examine our second example. We evaluate the time of execution and accuracy of approximation, as compared with the actual solution in the interval *[1, 2]*:

```
In [6]: def F(t, y): return 3.0*y**0.75 - 6.0*y/t
In [7]: def g(t): return (3.0*t**2.5 + 7)**4.0/(10000.0*t**6.)
In [8]: f = mpmath.odefun(F, 1.0, 1.0)
In [9]: t = np.linspace(1, 2, 1024); \
   ...: Y1 = np.vectorize(f)(t); \
   ...: Y2 = np.vectorize(g)(t)
In [9]: (np.abs(Y1-Y2)).max()
Out[9]: mpf('5.5511151231257827e-16')
```

Now let's address the example with the Latko-Volterra system. We compute the solutions and plot them for a time range of 0 to 10 units of time:

```
In [10]: def F(t, y): return [y[0]  - 0.1*y[0]*y[1],
    ....:                      0.075*y[0]*y[1] - 1.5*y[1]]
In [11]: f = mpmath.odefun(F, 0.0, [10.0, 5.0])
In [12]: T = [10.0*x/1023. for x in range(1024)]
    ....: X = [f(10.0*x/1023.)[0] for x in range(1024)]; \
    ....: Y = [f(10.0*x/1023.)[1] for x in range(1024)]
In [13]: plt.plot(T, X, 'r--', linewidth=2.0, label='predator'); \
    ....: plt.plot(T, Y, 'b-',  linewidth=2.0, label='prey'); \
    ....: plt.legend(loc=9); \
    ....: plt.grid(); \
    ....: plt.show()
```

This last command presents us with the following graph. The dotted line represents the amount of predators with respect to time, and the solid line represents the prey. First note the periodicity of the solution. Also, note the behavior of both functions—when the number of predators is high, the amount of prey is low. At that point, predators have a harder time finding food and their numbers start decreasing, while that of the prey starts rising:

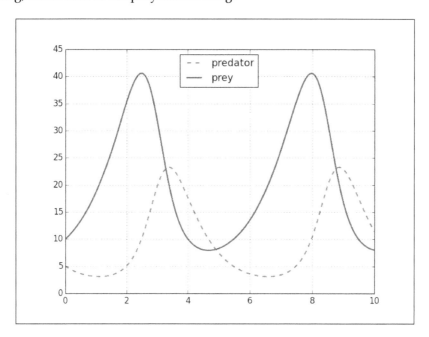

Discrete-variable methods

In discrete-variable methods, we are concerned with finding approximations to the solutions, but only at a discrete set of points in the domain. These points could be predetermined before solving, or we could generate them dynamically as part of the integration, to better suit the properties of the functions involved. This is especially useful when the solutions present singularities, for example, once a discrete set of points have been generated, we can compute a nice analytic approximation to the solutions, by a simple process of interpolation.

We have two schema for discrete-variable methods:

• **One-step methods**: The value of the solution at one point is computed solely from information on the previous point. Classical exponents of this scheme are, for instance, Euler's method, improved Euler's method, any second-order two-stage method, or any Runge-Kutta method.

- **Multistep methods**: The value of the solution at one point depends on the knowledge of several previous points. The best known algorithms in this category are the Adams-Bashford method, the Adams-Moulton method, any backward-difference method, or any predictor-corrector method.

In the module `scipy.integrate`, we have the common interface class `ode`, that will perform an approximation to the solution of equations/systems with a chosen numerical method. The way to work with this class is very different compared to what we are used to, and it deserves a careful explanation:

1. Once a right-hand side of a differential equation/system is produced, say $y' = f(t,y)$, the process starts by creating an instance of a solver. We do so by issuing `ode(f)`. If we have a description of the Jacobian of the right-hand side with respect to the y variables, we could include it in the creation of the solver `ode(f, jac)`.

2. If extra parameters need to be fed to either the function `f` or its Jacobian, we do so with `.set_f_params(*args)` or `.set_jac_params(*args)`, respectively.

3. The initial values of the problem are indicated with the attribute `.set_initial_value(y[, t])`.

4. It is time to choose a numerical scheme. We do so by setting the attribute `.set_integrator(name, **params)`. If necessary, we can provide further information to the chosen method, by using the optional parameter.

5. Finally, we compute the actual solution of the initial value problem. We usually accomplish this by playing with several attributes within a loop:

   - `.integrate(t[, step, relax])` will compute the value of the solution `y(t)` at the provided time `t`.
   - Retrieval of the last step in the computations can always be obtained with the `attributes.t` (for the time variable) and `.y` (for the solution)
   - To check for the success of the computation, we have the attribute `.successful()`.
   - Some integration methods accept a flag function `solout_func(t, y)`, that gets called after each successful step. This is accomplished with `.set_solout(solout_func)`.

One-step methods

The only one-step methods coded in the SciPy stack are two implementations of Runge-Kutta, designed by Dormand and Prince, and written by Hairer and Wanner for the module `scipy.integrate`:

- Explicit Runge-Kutta method of order (4)5. We access it with `method='dopri5'`.

- Explicit Runge-Kutta method of order 8(5,3). We call it with `method='dop853'`.

Let's run through our examples. With the first one, we will solve the differential equation by issuing `dopri5`, on a set of 10 nodes given by zeros of the Chebyshev polynomial in the interval *[0, 1]*:

```
In [1]: import numpy as np,import matplotlib.pyplot as plt; \
   ...: from scipy.integrate import ode; \
   ...: from scipy.special import t_roots
In [2]: def F(t, y): return y
In [3]: solver = ode(F)          # solver created
In [4]: solver.set_initial_value(1.0, 0.0)  # y(0) = 1
Out[4]: <scipy.integrate._ode.ode at 0x1038d3a50>
In [5]: solver.set_integrator('dopri5')
Out[5]: <scipy.integrate._ode.ode at 0x1038d3a50>
In [6]: solver.t, solver.y
Out[6]: (0.0, array([ 1.]))
In [7]: nodes = t_roots(10)[0]; \
   ...: nodes = (nodes + 1.0) * 0.5
In [8]: for k in range(10):
   ...:       if solver.successful():
   ...:            t = nodes[k]
   ...:            solver.integrate(t)
   ...:            print "{0},{1},{2}".format(t, solver.y[0], np.exp(t))
   ...:
0.00615582970243, 1.00617481576, 1.00617481576
0.0544967379058, 1.05600903161, 1.05600903161
0.146446609407, 1.15771311835, 1.15771311818
0.27300475013, 1.31390648743, 1.31390648604
0.42178276748, 1.52467728436, 1.52467727922
0.57821723252, 1.78285718759, 1.78285717608
0.72699524987, 2.06885488518, 2.06885486703
0.853553390593, 2.34797534217, 2.34797531943
```

```
0.945503262094, 2.57410852921, 2.5741085039
0.993844170298, 2.7015999731, 2.70159994653
```

It is possible to fine-tune the algorithm by providing different tolerances, restriction of number of steps, and other stabilizing constants. For a detailed description of the different parameters, refer to the official documentation at http://docs.scipy.org/doc/scipy/reference/generated/scipy.integrate.ode.html, or simply request the manual page from within your Python session with the following code:

```
>>> help(ode)
```

In the example of the Bernoulli equation, we will again gather the roots of a Chebyshev polynomial as nodes, but this time we will collect the solutions and construct a piecewise polynomial interpolation with them, to compare the results with the true solution visually. In this case, we employ Runge-Kutta of order 8(5,3):

```
In [9]: def bernoulli(t, y): return 3*y**(0.75) - 6.0*y/t
In [10]: def G(t):
   ....:     return (3.0*t**(2.5) + 7.0)**4.0 / (10000.0*t**6.0)
In [11]: solver = ode(bernoulli); \
   ....: solver.set_initial_value(1.0, 1.0); \
   ....: solver.set_integrator('dop853')
Out[11]: <scipy.integrate._ode.ode at 0x104667f50>
In [12]: T = np.linspace(1, 2, 1024); \
   ....: nodes = t_roots(10)[0]; \
   ....: nodes = 1.5 + 0.5 * nodes; \
   ....: solution = []
In [13]: for k in range(10):
   ....:     if solver.successful():
   ....:         solver.integrate(nodes[k])
   ....:         solution += [solver.y[0]]
   ....:
In [14]: from scipy.interpolate import PchipInterpolator
In [15]: interpolant = PchipInterpolator(nodes, solution)
In [16]: plt.plot(T, interpolant(T), 'r--', \
   ....:             linewidth=2.0, label='approx.'); \
   ....: plt.plot(T, G(T), 'b-', label='true soln.'); \
   ....: plt.grid(); \
   ....: plt.legend(); \
   ....: plt.show()
```

This presents the following enlightening diagram. The computed solution (dotted line) very accurately resembles the true solution (solid line) between times t=1 and t=2:

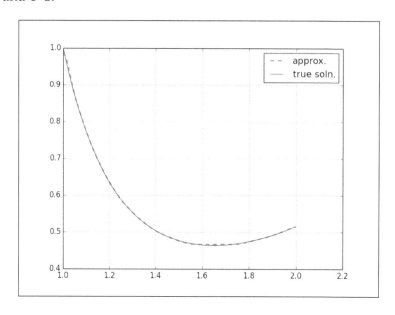

The Lotka-Volterra system gets solved in the same fashion. In the following example, we will choose a set of equally spaced nodes in one cycle — the interval from 0 to $2\pi/\sqrt{1.5}$ (approximately 5.13):

```
In [16]: def volterra(t, y):
    ....:         return [y[0]   - 0.1*y[0]*y[1],
    ....:                  0.075*y[0]*y[1] - 1.5*y[1]]
In [17]: solver = ode(volterra); \
    ....: solver.set_initial_value([10.0, 5.0], 0.0); \
    ....: solver.set_integrator('dop853')
Out[17]: <scipy.integrate._ode.ode at 0x10461e390>
In [18]: prey = []; \
    ....: predator = []
In [19]: while (solver.t < 5.13 and solver.successful()):
    ....:         solver.integrate(solver.t + 0.01)
    ....:         prey += [solver.y[0]]
```

```
    ....:        predator += [solver.y[1]]
    ....:
In [20]: plt.plot(prey, predator); \
    ....: plt.grid(); \
    ....: plt.xlabel('number of prey'); \
    ....: plt.ylabel('number of predators'); \
    ....: plt.show()
```

This presents us with a phase portrait of the system. The graph represents a curve where, for each unit of time, the x coordinate represents the number of prey, and the corresponding y coordinate represents the number of predators. Due to the periodicity of the solutions, the phase portrait is a closed curve. As it happened with the simple plot of both solutions, the phase portrait also illustrates how, for example, when the number of predators raises over 20 units, the number of predators is usually small (less than 20 units). When the number of predators goes below 20, the number of prey rises slowly to over 40 units:

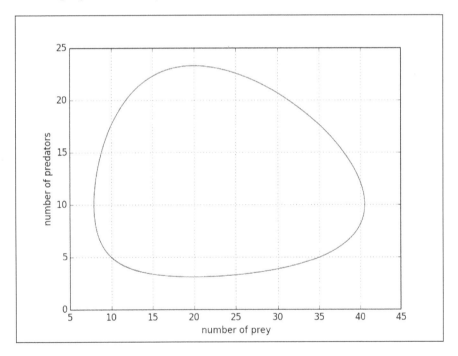

Two-step methods

We have two different choices here—an Adams-Moulton method (suitable for non-stiff equations) and a backward-difference method (designed for stiff equations). For each of these two numerical methods, we have two different implementations, depending on the background Fortran routine used to compute the solutions. The options are as follows:

- *VODE: In the Fortran library ODE, we have the routines VODE and ZVODE (for a real and complex-valued solution of initial value problems, respectively). To access the numerical method of Adams-Moulton, for example, for a real-valued problem, we issue the attribute .set_integrator('vode', method='adams'). To access backward-differences, we issue .set_integrator('vode', method='BDF').

- LSODA: This other implementation wraps different routines from the Fortran library ODEPACK. The calls are exactly as in the previous case, substituting 'vode' or 'zvode' with 'lsoda' instead.

 For more information about the netlib libraries ODE and ODEPACK, refer to http://www.netlib.org/ode/ and http://www.netlib.org/odepack/, respectively.

These numerical methods are designed for large problems. For smaller tasks (non-stiff one-dimensional equations with small set of nodes), Runge-Kutta should be used instead. The second example illustrates this point: We apply BDF from VODE, and compare the solutions obtained with both Runge-Kutta from before, to the actual solution. Note how dop853 outperforms BDF in this simple case:

```
In [21]: solver = ode(bernoulli); \
    ....: solver.set_initial_value(1.0, 1.0); \
    ....: solver.set_integrator('vode', method='BDF')
Out[21]: <scipy.integrate._ode.ode at 0x1038d2990>
In [22]: nodes = t_roots(10)[0]; \
    ....: nodes = 1.5 + 0.5 * nodes; \
    ....: solution2 = []
In [23]: for k in range(10):
    ....:     if solver.successful():
    ....:         solver.integrate(nodes[k])
    ....:         solution2 += [solver.y[0]]
    ....:
```

```
In [24]: for k in range(10):
    ....:     true = G(nodes[k])
    ....:     dop853 = solution[k]
    ....:     vode = solution2[k]
    ....:     print "{0},{1},{2}".format(true, dop853, vode)
    ....:
0.981854827818, 0.981854827818, 0.981855789349
0.859270248468, 0.859270248468, 0.859270080689
0.698456572663, 0.698456572663, 0.698458875953
0.570963875566, 0.570963875559, 0.57096210196
0.49654019444, 0.496540194433, 0.496537599383
0.466684337536, 0.466684337531, 0.466681706374
0.466700776675, 0.46670077667, 0.466699552511
0.482536531899, 0.482536531895, 0.482537207918
0.501698475572, 0.501698475568, 0.501699366611
0.514159743133, 0.514159743128, 0.514160809167
```

To finish this chapter, we use LSODA on our Lotka-Volterra system, including information of the Jacobian:

```
In [25]: def jacF(t, y):
    ....:     output = np.zeros((2,2))
    ....:     output[0,0] = 1.0 - 0.1*y[1]
    ....:     output[0,1] = -0.1*y[0]
    ....:     output[1,0] = 0.075*y[1]
    ....:     output[1,1] = 0.075*y[0] - 1.5
    ....:     return output
In [26]: solver = ode(volterra, jacF); \
    ....: solver.set_initial_value([10.0, 5.0], 0.0); \
    ....: solver.set_integrator('lsoda', method='adams',
    ....:                       with_jacobian=True)
In [27]: prey2 = []; \
    ....: predator2  = []
In [28]: while (solver.t < 5.13 and solver.successful()):
    ....:     solver.integrate(solver.t + 0.01)
    ....:     prey2 += [solver.y[0]]
    ....:     predator2 += [solver.y[1]]
```

We leave this as an exercise to compare the results of solving this last system with this method against the previous Runge-Kutta process.

Summary

In this short chapter, we have mastered all the symbolic and numerical techniques to solve differential equations/systems and related initial value problems.

In the next chapter, we will explore what resources we have in the SciPy stack, to address problems in computational geometry.

6

Computational Geometry

Computational geometry is a field of mathematics that seeks the development of efficient algorithms to solve problems described in terms of basic geometrical objects. We differentiate between **combinatorial computational geometry** and **numerical computational geometry**.

Combinatorial computational geometry deals with the interaction of basic geometrical objects — points, segments, lines, polygons, and polyhedra. In this setting, we have three categories of problems:

- **Static problems**: The construction of a known target object is required from a set of input geometric objects
- **Geometric query problems**: Given a set of known objects (the search space) and a sought property (the query), these problems deal with the search of objects that satisfy the query
- **Dynamic problems**: This is similar to the problems from the previous two categories, with the added challenge that the input is not known in advance, and objects are inserted or deleted between queries/constructions

Numerical computational geometry deals mostly with the representation of objects in space described by means of curves, surfaces, and regions in space bounded by these.

Before we proceed to the development and analysis of the different algorithms in those two settings, it pays to explore the basic background — Plane geometry.

Plane geometry

The geometry module of the SymPy library covers basic geometry capabilities. Rather than giving an academic description of all objects and properties in that module, we discover the most useful ones through a series of small self-explanatory Python sessions.

We start with the concepts of *point* and *segment*. The aim is to illustrate how easily we can check for collinearity, compute lengths, midpoints, or slopes of segments, for example. We also show how to quickly compute the angle between two segments, as well as decide whether a given point belongs to a segment or not. The following diagram illustrates an example, which we will follow up with the code:

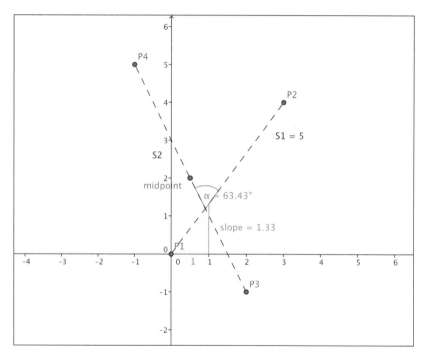

```
In [1]: from sympy.geometry import Point, Segment, Line, \
   ...:                           Circle, Triangle, Curve
In [2]: P1 = Point(0, 0); \
   ...: P2 = Point(3, 4); \
   ...: P3 = Point(2, -1); \
   ...: P4 = Point(-1, 5)
In [3]: statement = Point.is_collinear(P1, P2, P3); \
   ...: print "Are P1, P2, P3 collinear?," statement
Are P1, P2, P3 collinear? False
```

```
In [4]: S1 = Segment(P1, P2); \
   ...: S2 = Segment(P3, P4)
In [5]: print "Length of S1:", S1.length
Length of S1: 5
In [6]: print "Midpoint of S2:", S2.midpoint
Midpoint of S2: Point(1/2, 2)
In [7]: print "Slope of S1", S1.slope
Slope of S1: 4/3
In [8]: print "Intersection of S1 and S2:", S1.intersection(S2)
Intersection of S1 and S2: [Point(9/10, 6/5)]
In [9]: print "Angle between S1, S2:", Segment.angle_between(S1, S2)
Angle between S1, S2: acos(-sqrt(5)/5)
In [10]: print "Does S1 contain P3?", S1.contains(P3)
Does S1 contain P3? False
```

The next logical geometrical concept is the *line*. We can perform more interesting operations with lines, and to that effect, we have a few more constructors. We can find their equations; compute the distance between a point and a line, and many other operations:

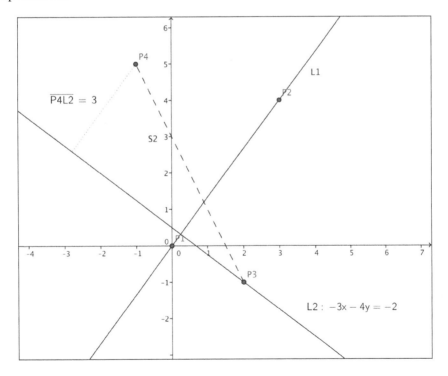

```
In [11]: L1 = Line(P1, P2)
In [12]: L2 = L1.perpendicular_line(P3) #perpendicular line to L1
In [13]: print "Parametric equation of L2:", L2.arbitrary_point()
Parametric equation of L2: Point(4*t + 2, -3*t - 1)
In [14]: print "Algebraic equation of L2:", L2.equation()
Algebraic equation of L2: 3*x + 4*y - 2
In [15]: print "Does L2 contain P4?", L2.contains(P4)
Does L2 contain P4? False
In [16]: print "Distance from P4 to L2:", L2.distance(P4)
Distance from P4 to L2: 3
In [17]: print "Is L2 parallel with S2?", L1.is_parallel(S2)
Is L2 parallel with S2? False
```

The next geometrical concept we are to explore is the *circle*. We can define a circle by its center and radius, or by three points on it. We can easily compute all of its properties, as shown in the following diagram:

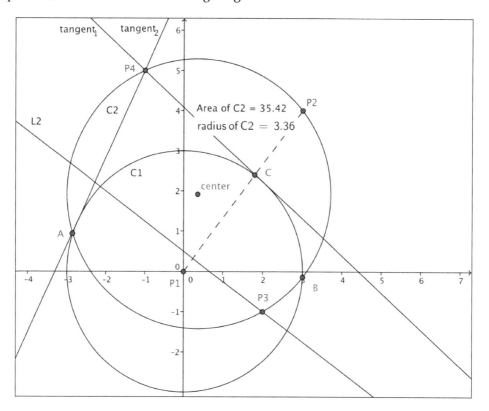

```
In [18]: C1 = Circle(P1, 3); \
   ....: C2 = Circle(P2, P3, P4)
In [19]: print "Area of C2:", C2.area
Area of C2: 1105*pi/98
In [20]: print "Radius of C2:", C2.radius
Radius of C2: sqrt(2210)/14
In [21]: print "Algebraic equation of C2:", C2.equation()
Algebraic equation of C2: (x - 5/14)**2 + (y - 27/14)**2 - 1105/98
In [22]: print "Center of C2:", C2.center
Center of C2: Point(5/14, 27/14)
In [23]: print "Circumference of C2:", C2.circumference
Circumference of C2: sqrt(2210)*pi/7
```

Computing intersections with other objects, checking whether a line is tangent to a circle, or finding the tangent lines through a exterior point, are simple tasks too:

```
In [24]: print "Intersection of C1 and C2:\n", C2.intersection(C1)
Intersection of C1 and C2:
[Point(55/754 + 27*sqrt(6665)/754, -5*sqrt(6665)/754 + 297/754),
 Point(-27*sqrt(6665)/754 + 55/754, 297/754 + 5*sqrt(6665)/754)]
In [25]: print "Intersection of S1 and C2:\n", C2.intersection(S1)
Intersection of S1 and C2:
[Point(3, 4)]
In [26]: print "Is L2 tangent to C2?", C2.is_tangent(L2)
Is L2 tangent to C2? False
In [27]: print "Tangent lines to C1 through P4:\n", \
              C1.tangent_lines(P4)
Tangent lines to C1 through P4:
[Line(Point(-1, 5),
      Point(-9/26 + 15*sqrt(17)/26, 3*sqrt(17)/26 + 45/26)),
 Line(Point(-1, 5),
      Point(-15*sqrt(17)/26 - 9/26, -3*sqrt(17)/26 + 45/26))]
```

The *triangle* is a very useful basic geometric concept. Reliable manipulation of these objects is at the core of computational geometry. We need robust and fast algorithms to manipulate and extract information from them. Let's first show the definition of one, together with a series of queries to describe its properties:

```
In [28]: T = Triangle(P1, P2, P3)
In [29]: print "Signed area of T:", T.area
Signed area of T: -11/2
In [30]: print "Angles of T:\n", T.angles
Angles of T:
{Point(3, 4): acos(23*sqrt(26)/130),
```

```
  Point(2, -1): acos(3*sqrt(130)/130),
  Point(0, 0): acos(2*sqrt(5)/25)}
In [31]: print "Sides of T:\n", T.sides
Sides of T:
[Segment(Point(0, 0), Point(3, 4)),
 Segment(Point(2, -1), Point(3, 4)),
 Segment(Point(0, 0), Point(2, -1))]
In [32]: print "Perimeter of T:", T.perimeter
Perimeter of T: sqrt(5) + 5 + sqrt(26)
In [33]: print "Is T a right triangle?", T.is_right()
Is T a right triangle? False
In [34]: print "Is T equilateral?", T.is_equilateral()
Is T equilateral? False
In [35]: print "Is T scalene?", T.is_scalene()
Is T scalene? True
In [36]: print "Is T isosceles?", T.is_isosceles()
Is T isosceles? False
```

Next, note how easily we can obtain representation of the different segments, centers, and circles associated with triangles, as well as the medial triangle (the triangle with vertices at the midpoints of the segments):

```
In [37]: T.altitudes
Out[37]:
{Point(0, 0) : Segment(Point(0, 0), Point(55/26, -11/26)),
 Point(2, -1): Segment(Point(6/25, 8/25), Point(2, -1)),
 Point(3, 4) : Segment(Point(4/5, -2/5), Point(3, 4))}
In [38]: T.orthocenter     # Intersection of the altitudes
Out[38]:
Point((3*sqrt(5) + 10)/(sqrt(5) + 5 + sqrt(26)),
      (-5 + 4*sqrt(5))/(sqrt(5) + 5 + sqrt(26)))
In [39]: T.bisectors()     # Angle bisectors
Out[39]:
{Point(0, 0) : Segment(Point(0, 0),
                       Point(sqrt(5)/4 + 7/4, -9/4 + 5*sqrt(5)/4)),
 Point(2, -1): Segment(Point(3*sqrt(5)/(sqrt(5) + sqrt(26)),
                             4*sqrt(5)/(sqrt(5) + sqrt(26))),
          Point(2, -1)),
 Point(3, 4) : Segment(Point(-50 + 10*sqrt(26), -5*sqrt(26) + 25),
                       Point(3, 4))}
In [40]: T.incenter     # Intersection of angle bisectors
Out[40]:
Point((3*sqrt(5) + 10)/(sqrt(5) + 5 + sqrt(26)),
      (-5 + 4*sqrt(5))/(sqrt(5) + 5 + sqrt(26)))
In [41]: T.incircle
```

```
Out[41]:
Circle(Point((3*sqrt(5) + 10)/(sqrt(5) + 5 + sqrt(26)),
             (-5 + 4*sqrt(5))/(sqrt(5) + 5 + sqrt(26))),
       -11/(sqrt(5) + 5 + sqrt(26)))
In [42]: T.inradius
Out[42]: -11/(sqrt(5) + 5 + sqrt(26))
In [43]: T.medians
Out[43]:
{Point(0, 0) : Segment(Point(0, 0), Point(5/2, 3/2)),
 Point(2, -1): Segment(Point(3/2, 2), Point(2, -1)),
 Point(3, 4) : Segment(Point(1, -1/2), Point(3, 4))}
In [44]: T.centroid      # Intersection of the medians
Out[44]: Point(5/3, 1)
In [45]: T.circumcenter   # Intersection of perpendicular bisectors
Out[45]: Point(45/22, 35/22)
In [46]: T.circumcircle
Out[46]: Circle(Point(45/22, 35/22), 5*sqrt(130)/22)
In [47]: T.circumradius
Out[47]: 5*sqrt(130)/22
In [48]: T.medial
Out[48]: Triangle(Point(3/2, 2), Point(5/2, 3/2), Point(1, -1/2))
```

Here are some other interesting operations with triangles:

- Intersection with other objects
- Computation of the minimum distance from a point to each of the segments
- Checking whether two triangles are similar

```
In [49]: T.intersection(C1)
Out[49]: [Point(9/5, 12/5),
          Point(sqrt(113)/26 + 55/26, -11/26 + 5*sqrt(113)/26)]
In [50]: T.distance(T.circumcenter)
Out[50]: sqrt(26)/11
In [51]: T.is_similar(Triangle(P1, P2, P4))
Out[51]: False
```

The other basic geometrical objects currently coded in the geometry module are:

- `LinearEntity`: This is a superclass with three subclasses: `Segment`, `Line`, and `Ray`. The class `LinearEntity` enjoys the following basic methods:
    - `are_concurrent(o1, o2, ..., on)`
    - `are_parallel(o1, o2)`
    - `are_perpendicular(o1, o2)`

- ○ `parallel_line(self, Point)`
- ○ `perpendicular_line(self, Point)`
- ○ `perpendicular_segment(self, Point)`

- `Ellipse`: This is an object with a center, together with horizontal and vertical radii. `Circle` is, as a matter of fact, a subclass of `Ellipse` with both radii equal.

- `Polygon`: This is a superclass that we can instantiate by listing a set of vertices. `Triangles` are a subclass of `Polygon`, for example. The basic methods of Polygon are:

  - ○ `area`
  - ○ `perimeter`
  - ○ `centroid`
  - ○ `sides`
  - ○ `vertices`

- `RegularPolygon`. This is a subclass of `Polygon`, with extra attributes:

  - ○ `apothem`
  - ○ `center`
  - ○ `circumcircle`
  - ○ `exterior_angle`
  - ○ `incircle`
  - ○ `interior_angle`
  - ○ `radius`

 For more information about this module, refer to the official SymPy documentation at `http://docs.sympy.org/latest/modules/geometry/index.html`.

There is also a nonbasic geometric object—a *curve*, which we define by providing parametric equations, together with the interval of definition of the parameter. It currently does not have many useful methods, other than those describing its constructors. Let's illustrate how to deal with these objects. For example, a three-quarters arc of an ellipse could be coded as follows:

```
In [52]: from sympy import var, pi, sin, cos
In [53]: var('t', real=True)
In [54]: Arc = Curve((3*cos(t), 4*sin(t)), (t, 0, 3*pi/4))
```

To end the exposition on basic objects from the geometry module in the SymPy library, we must mention that we can apply any basic affine transformations to any of the previous objects. This is done by combination of the methods `reflect`, `rotate`, `translate`, and `scale`:

```
In [55]: T.reflect(L1)
Out[55]: Triangle(Point(0, 0), Point(3, 4), Point(-38/25, 41/25))
In [56]: T.rotate(pi/2, P2)
Out[56]: Triangle(Point(7, 1), Point(3, 4), Point(8, 3))
In [57]: T.translate(5,4)
Out[57]: Triangle(Point(5, 4), Point(8, 8), Point(7, 3))
In [58]: T.scale(9)
Out[58]: Triangle(Point(0, 0), Point(27, 4), Point(18, -1))
In [59]: Arc.rotate(pi/2, P3).translate(pi,pi).scale(0.5)
Out[59]:
Curve((-2.0*sin(t) + 0.5 + 0.5*pi, 3*cos(t) - 3 + pi),
      (t, 0, 3*pi/4))
```

With these basic definitions and operations, we are ready to address more complex situations. Let's explore these new challenges next.

Combinatorial computational geometry

Also called algorithmic geometry, the applications of this field are plenty. In robotics, it is used to solve visibility problems, and motion planning, for instance. Similar applications are employed to design route planning or search algorithms in **geographic information systems** (**GIS**).

Let's describe the different categories of problems, putting emphasis on the tools to solve them, which are available in the SciPy stack.

Static problems

The fundamental problems in this category are the following:

- **Convex hulls**: Given a set of points in space, find the smallest convex polyhedron containing them.

- **Voronoi diagrams**: Given a set of points in space (the seeds), compute a partition in regions consisting of all points closer to each seed.

- **Triangulations**: Partition the plane with triangles in a way that two triangles are either disjoint, or otherwise they share an edge or a vertex. There are different triangulations depending on the input objects or constraints on the properties of the triangles.

- **Shortest paths**: Given a set of obstacles in a space and two points, find the shortest path between the points that does not intersect any of the obstacles.

The problems of computation of convex hulls, basic triangulations, and Voronoi diagrams are intimately linked. The theory that explains this beautiful topic is explained in detail in a monograph in computer science titled *Computational Geometry*, written by Franco Preparata and Michael Shamos. It was published by Springer-Verlag in 1985.

Convex hulls

While it is possible to compute the convex hull of a reasonably large set of points in the plane through the geometry module of the library SymPy, this is not recommended. A much faster and reliable code is available in the module `scipy.spatial` through the class `ConvexHull`, which implements a wrapper to the routine `qconvex`, from the Qhull libraries (http://www.qhull.org/). This routine also allows the computation of convex hulls in higher dimensions. Let's compare both methods, with the famous Lake Superior polygon, `superior.poly`.

The poly files represent planar straight line graphs—a simple list of vertices and edges, together with information about holes and concavities, in some cases. The running example can be downloaded from https://github.com/blancosilva/Mastering-Scipy/blob/master/chapter6/superior.poly.

This contains a polygonal description of the coastline of Lake Superior, with 7 holes (for the islands), 518 vertices, and 518 edges.

For a complete description of the poly format, refer to http://www.cs.cmu.edu/~quake/triangle.poly.html. With that information, we can write a simple reader without much effort.

Following is an example.

```
# part of file chapter6.py
from numpy import array
def read_poly(file_name):
    """
    Simple poly-file reader, that creates a python
        dictionary
    with information about vertices, edges and holes.
    It assumes that vertices have no attributes or
        boundary markers.
    It assumes that edges have no boundary markers.
    No regional attributes or area constraints are
```

```
        parsed.
    """
    output = {'vertices': None,
                 'holes': None,
                 'segments': None}
    # open file and store lines in a list
    file = open(file_name, 'r')
    lines = file.readlines()
    file.close()
    lines = [x.strip('\n').split() for x in lines]
    # Store vertices
    vertices= []
    N_vertices,dimension,attr,bdry_markers = [int(x) for x in lines[0]]
    # We assume attr = bdrt_markers = 0
    for k in range(N_vertices):
      label,x,y = [items for items in lines[k+1]]
      vertices.append([float(x), float(y)])
    output['vertices']=array(vertices)
    # Store segments
    segments = []
    N_segments,bdry_markers = [int(x) for x in lines[N_vertices+1]]
    for k in range(N_segments):
      label,pointer_1,pointer_2 = [items for items in lines[N_
vertices+k+2]]
      segments.append([int(pointer_1)-1, int(pointer_2)-1])
    output['segments'] = array(segments)
    # Store holes
    N_holes = int(lines[N_segments+N_vertices+2][0])
    holes = []
    for k in range(N_holes):
      label,x,y = [items for items in lines[N_segments + N_vertices + 3 +
k]]
      holes.append([float(x), float(y)])
    output['holes'] = array(holes)
    return output
```

Notice that loading each vertex as `Point`, as well as computing the convex hull with that structure, requires far too many resources and too much time. Note the difference:

```
In [1]: import numpy as np, matplotlib.pyplot as plt; \
   ...: from sympy.geometry import Point, convex_hull; \
   ...: from scipy.spatial import ConvexHull; \
   ...: from chapter6 import read_poly
In [2]: lake_superior = read_poly("superior.poly"); \
   ...: vertices_ls = lake_superior['vertices']
```

```
In [3]: %time hull = ConvexHull(vertices_ls)
CPU times: user 1.59 ms, sys: 372 µs, total: 1.96 ms
Wall time: 1.46 ms
In [4]: vertices_sympy = [Point(x) for x in vertices_ls]
In [5]: %time convex_hull(*vertices_sympy)
CPU times: user 168 ms, sys: 54.5 ms, total: 223 ms
Wall time: 180 ms
Out[5]:
Polygon(Point(1/10, -629607/1000000),
        Point(102293/1000000, -635353/1000000),
        Point(2773/25000, -643967/1000000),
        Point(222987/1000000, -665233/1000000),
        Point(8283/12500, -34727/50000),
        Point(886787/1000000, -1373/2000),
        Point(890227/1000000, -6819/10000),
        Point(9/10, -30819/50000),
        Point(842533/1000000, -458913/1000000),
        Point(683333/1000000, -17141/50000),
        Point(16911/25000, -340427/1000000),
        Point(654027/1000000, -333047/1000000),
        Point(522413/1000000, -15273/50000),
        Point(498853/1000000, -307193/1000000),
        Point(5977/25000, -25733/50000),
        Point(273/2500, -619833/1000000))
```

Let us produce a diagram with the solution, using the computations of
`scipy.spatial.ConvexHull`:

 Plotting a set of vertices together with its convex hull in two dimensions
(once computed with `ConvexHull`) is also possible with the simple
command `convex_hull_plot_2d`. It requires `matplotlib.pyplot`.

```
In [5]: plt.figure(); \
   ...: plt.xlim(vertices_ls[:,0].min()-0.01,
               vertices_ls[:,0].max()+0.01); \
   ...: plt.ylim(vertices_ls[:,1].min()-0.01,
               vertices_ls[:,1].max()+0.01); \
   ...: plt.axis('off'); \
   ...: plt.axes().set_aspect('equal'); \
   ...: plt.plot(vertices_ls[:,0], vertices_ls[:,1], 'b.')
Out[5]: [<matplotlib.lines.Line2D at 0x10ee3ab10>]
In [6]: for simplex in hull.simplices:
   ...:     plt.plot(vertices_ls[simplex, 0],
   ...:              vertices_ls[simplex, 1], 'r-')
In [7]: plt.show()
```

This plots the following image:

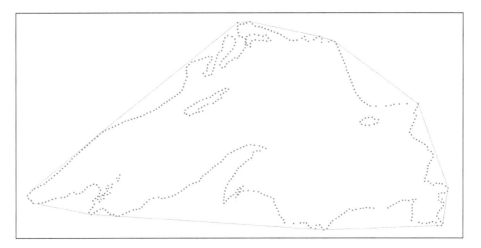

To modify the output of ConvexHull, we are allowed to pass all required qconvex controls through the parameter qhull_options. For a list of all qconvex controls and other output options, consult the Qhull manual at http://www.qhull.org/html/index.htm. In this chapter, we are content with showing the results obtained with the default controls qhull_options='Qx Qt' if the dimension of the points is greater than four, and qhull_options='Qt' otherwise.

Let's now illustrate a few advanced uses of ConvexHull. First, the computation of the convex hull of a random set of points in the 3D space. For visualization, we will use the mayavi libraries:

```
In [8]: points = np.random.rand(320, 3)
In [9]: hull = ConvexHull(points)
In [10]: X = hull.points[:, 0]; \
    ....: Y = hull.points[:, 1]; \
    ....: Z = hull.points[:, 2]
In [11]: from mayavi import mlab
In [12]: mlab.triangular_mesh(X, Y, X, hull.simplices,
    ....:                     colormap='gray', opacity=0.5,
    ....:                     representation='wireframe')
```

This plots the following image:

Voronoi diagrams

Computing the Voronoi diagram of a set of vertices (our seeds) can be done with the class `Voronoi` (and its companion `voronoi_plot_2d` for visualization) from the module `scipy.spatial`. This class implements a wrapper to the routine `qvoronoi` from the `Qhull` libraries, with the following default controls `qhull_option='Qbb Qc Qz Qx'` if the dimension of the points is greater than four, and `qhull_options='Qbb Qc Qz'` otherwise. For the computation of the furthest-site Voronoi diagram, instead of the nearest-site, we would add the extra control `'Qu'`.

Let's work a simple example with the usual Voronoi diagram:

```
In [13]: from scipy.spatial import Voronoi, voronoi_plot_2d
In [14]: vor = Voronoi(vertices_ls)
```

To understand the output, it is very illustrative to replicate the diagram that we obtain by restricting the visualization obtained by `voronoi_plot_2d` in a small window, centered somewhere along the north shore of Lake Superior:

```
In [15]: ax = plt.subplot(111, aspect='equal'); \
   ....: voronoi_plot_2d(vor, ax=ax); \
   ....: plt.xlim( 0.45,  0.50); \
   ....: plt.ylim(-0.40, -0.35); \
   ....: plt.show()
```

This plots the following image:

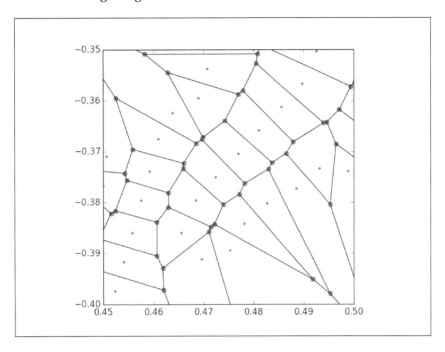

- The small dots are the original seeds with *x* coordinates between `0.45` and `0.50`, and *y* coordinates between `-0.40` and `-0.35`. We access those values either from the original list `vertices_ls` or from `vor.points`.

- The plane gets partitioned into different regions (the Voronoi cells), one for each seed. These regions contain all the points in the plane which are closest to its seed. Each region receives an index, which is not necessarily the same index as the index of its seed in the `vor.points` list. To access the corresponding region to a given seed, we use `vor.point_region`:

```
In [16]: vor.point_region
Out[16]:
array([  0,  22,  24,  21,  92,  89,  91,  98,  97,  26, 218,
       219, 220, 217, 336, 224, 334, 332, 335, 324, 226, 231,
       230, 453, 500, 454, 235, 234, 333, 236, 341, 340,  93,
       ...
       199,  81,  18,  17, 205, 290,  77, 503, 469, 473, 443,
       373, 376, 366, 370, 369, 210, 251, 367, 368, 377, 472,
       504, 506, 502, 354, 353,  54,  42,  43, 350, 417, 414,
       415, 418, 419, 425])
```

- Each Voronoi cell is defined by its delimiting vertices and edges (also known as ridges in Voronoi jargon). The list with the coordinates of the computed vertices of the Voronoi diagram can be obtained with `vor.vertices`. These vertices were represented as bigger dots in the previous image, and are easily identifiable because they are always at the intersection of at least two edges—while the seeds have no incoming edges:

```
In [17]: vor.vertices
Out[17]:
array([[ 0.88382749, -0.23508215],
       [ 0.10607886, -0.63051169],
       [ 0.03091439, -0.55536174],
       ...,
       [ 0.49834202, -0.62265786],
       [ 0.50247159, -0.61971784],
       [ 0.5028735 , -0.62003065]])
```

- For each of the regions, we can access the set of delimiting vertices with `vor.regions`. For instance, to obtain the coordinates of the vertices that delimit the region around the fourth seed, we could issue the following command:

```
In [18]: [vor.vertices[x] for x in
               vor.regions[vor.point_region[4]]]
Out[18]:
[array([ 0.13930793, -0.81205929]),
 array([ 0.11638   , -0.92111088]),
 array([ 0.11638   , -0.63657789]),
 array([ 0.11862537, -0.6303235 ]),
 array([ 0.12364332, -0.62893576]),
 array([ 0.12405738, -0.62891987])]
```

Care must be taken with the previous step—some of the vertices of the Voronoi cells are not actual vertices, but lie at infinity. When this is the case, they are identified with the index -1. In this situation, to provide an accurate representation of a ridge of these characteristics, we must use the knowledge of the two seeds whose contiguous Voronoi cells intersect on said ridge—since the ridge is perpendicular to the segment defined by those two seeds. We obtain the information about those seeds with `vor.ridge_points`:

```
In [19]: vor.ridge_points
Out[19]:
array([[  0,   1],
       [  0, 433],
       [  0, 434],
       ...,
```

```
[124, 118],
[118, 119],
[119, 122]])
```

The first entry of `vor.ridge_points` can be read as, there is a ridge perpendicular to both the first and second seeds.

There are other attributes of the object `vor` that we could use to inquire properties of the Voronoi diagram, but the ones we have described should be enough to replicate the previous diagram. We leave this as a nice exercise:

1. Gather the indices of the seeds from `vor.points` that have their *x* coordinates and *y* coordinates in the required window. Plot them.

2. For each of those seeds, gather information about the vertices of their corresponding Voronoi cells. Plot those vertices that are not at infinity with a different style to the seeds.

3. Gather information about the ridges of each relevant region, and plot them as simple thin segments. Some of the ridges cannot be represented by their two vertices. In that case, we use the information about the seeds that determine them.

Triangulations

A triangulation of a set of vertices in the plane is a division of the convex hull of the vertices into triangles, satisfying one important condition. Any two given triangles can be either one of the following:

- They must be disjoint
- They must intersect only at one common vertex
- They must share one common edge

These plain triangulations have not much computational value, since some of their triangles might be too skinny—this leads to uncomfortable rounding errors, computation or erroneous areas, centers, and so on. Among all possible triangulations, we always seek one where the properties of the triangles are somehow balanced.

With this purpose in mind, we have the **Delaunay** triangulation of a set of vertices. This triangulation satisfies an extra condition—none of the vertices lie in the interior of the circumcircle of any triangle. We refer to triangles with this property as Delaunay triangles.

For this simpler setting, in the module `scipy.spatial`, we have the class `Delaunay`, which implements a wrapper to the routine `qdelaunay` from the `Qhull` libraries, with the controls set exactly, as in the case of the Voronoi diagram:

```
In [20]: from scipy.spatial import Delaunay
In [21]: tri = Delaunay(vertices_ls)
In [22]: plt.figure()
   ....: plt.xlim(vertices_ls[:,0].min()-0.01,
                   vertices_ls[:,0].max()+0.01); \
   ....: plt.ylim(vertices_ls[:,1].min()-0.01,
                   vertices_ls[:,1].max()+0.01); \
   ....: plt.axes().set_aspect('equal'); \
   ....: plt.axis('off'); \
   ....: plt.triplot(vertices_ls[:,0], vertices_ls[:,1],
                     tri.simplices, 'k-'); \
   ....: plt.plot(vertices_ls[:,0], vertices_ls[:,1], 'r.'); \
   ....: plt.show()
```

This plots the following diagram:

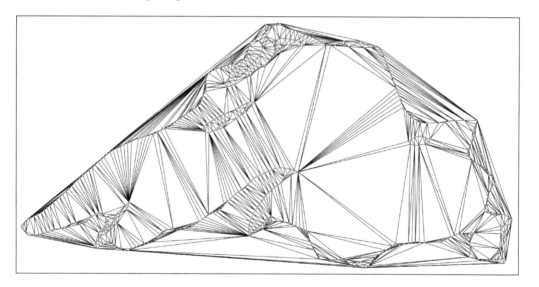

It is possible to generate triangulations with imposed edges too. Given a collection of vertices and edges, a **constrained Delaunay** triangulation is a division of the space into triangles with those prescribed features. The triangles in this triangulation are not necessarily Delaunay.

We can accomplish this extra condition sometimes by subdividing each of the imposed edges. We call this triangulation **conforming Delaunay**, and the new (artificial) vertices needed to subdivide the edges are called **Steiner points**.

A **constrained conforming Delaunay** triangulation of an imposed set of vertices and edges satisfies a few more conditions, usually setting thresholds on the values of angles or areas of the triangles. This is achieved by introducing a new set of Steiner points, which are allowed anywhere, not only on edges.

> To achieve these high-level triangulations, we need to step outside of the SciPy stack. We have a Python wrapper for the amazing implementation of mesh generators, `triangle`, by Richard Shewchuck (http://www.cs.cmu.edu/~quake/triangle.html). This wrapper, together with examples and other related functions, can be installed from prompt by issuing either `easy_install triangle`, or `pip install triangle`. For more information on this module, refer to the documentation online from its author, Dzhelil Rufat, at http://dzhelil.info/triangle/index.html.

Let's compute those different triangulations for our running example. We once again use the poly file with the features of Lake Superior, which we read into a dictionary with all the information about vertices, segments, and holes. The first example is that of the constrained Delaunay triangulation (`cndt`). We accomplish this task with the flag `p` (indicating that the source is a planar straight line graph, rather than a set of vertices):

```
In [23]: from triangle import triangulate, plot as tplot
In [24]: cndt = triangulate(lake_superior, 'p')
In [25]: ax = plt.subplot(111, aspect='equal'); \
   ....: tplot.plot(ax, **cndt); \
   ....: plt.show()
```

Note the improvement with respect to the previous diagram, as well as the absence of triangles outside of the original polygon:

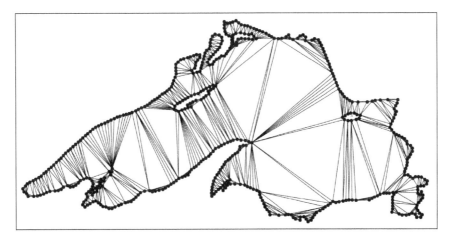

The next step is the computation of a conforming Delaunay triangulation (`cfdt`). We enforce Steiner points on some segments to ensure as many Delaunay triangles as possible. We achieve this with the extra flag `D`:

```
In [26]: cfdt = triangulate(lake_superior, 'pD')
```

Either slight or no improvement, with respect to the previous diagram, can be observed in this case. The real improvement arises when we further impose constraints in the values of minimum angles on triangles (with the flag `q`) or in the maximum values of the areas of triangles (with the flag `a`). For instance, if we require a constrained conforming Delaunay triangulation (`cncfdt`) in which all triangles have a minimum angle of at least 20 degrees, we issue the following command:

```
In [27]: cncfq20dt = triangulate(lake_superior, 'pq20D')
In [28]: ax = plt.subplot(111, aspect='equal'); \
    ....: tplot.plot(ax, **cncfq20dt); \
    ....: plt.show()
```

This presents us with a much more improved result, as seen in the following diagram:

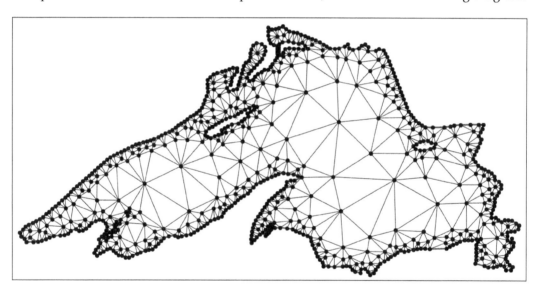

To conclude this section, we present a last example where we further impose a maximum area on triangles:

```
In [29]: cncfq20adt = triangulate(lake_superior, 'pq20a.001D')
In [30]: ax = plt.subplot(111, aspect='equal'); \
    ....: tplot.plot(ax, **cncfq20adt); \
    ....: plt.show()
```

The last (very satisfying) diagram is as follows:

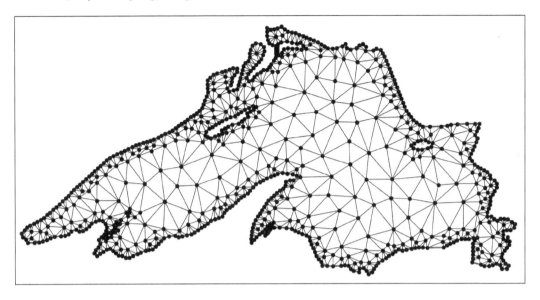

Shortest paths

We will use the previous example to introduce a special setting to the problem of shortest paths. We pick a location on the Northwest coast of the lake (say, the vertex indexed as 370 in the original poly file), and the goal is to compute the shortest path to the furthest Southeast location on the shore, at the bottom-right corner — this is the vertex indexed as 179 in the original poly file. By a path in this setting, we mean a chain of edges of the triangulation.

In the SciPy stack, we accomplish the computation of the shortest paths on a triangulation (and in some other similar geometries that can be coded by means of graphs) by relying on two modules:

- `scipy.sparse` is used to store a weighted-adjacency matrix G representing the triangulation. Each nonzero entry G[i,j] of this adjacency matrix is precisely the length of the edge from vertex i to vertex j.

- `scipy.sparse.csgraph` is the module that deals with compressed sparse graphs. This module contains routines to analyze, extract information, or manipulate graphs. Among these routines, we have several different algorithms to compute the shortest paths on a graph.

For more information on the module `scipy.sparse.csgraph`, refer to the online documentation at `http://docs.scipy.org/doc/scipy/reference/sparse.csgraph.html`.

For the theory and applications of Graph Theory, one of the best sources is the introductory book by Reinhard Diestel, *Graph Theory*, published by Springer-Verlag.

Let's illustrate this example with proper code. We start by collecting the indices of the vertices of all segments in the triangulation, and the lengths of these segments.

To compute the length of each segment, rather than creating, from scratch a routine that applies a reliable norm function over each item of the difference of two lists of related vertices, we use `minkowski_distance` from the module `scipy.spatial`.

```
In [31]: X = cncfq20adt['triangles'][:,0]; \
   ....: Y = cncfq20adt['triangles'][:,1]; \
   ....: Z = cncfq20adt['triangles'][:,2]
In [32]: Xvert = [cncfq20adt['vertices'][x] for x in X]; \
   ....: Yvert = [cncfq20adt['vertices'][y] for y in Y]; \
   ....: Zvert = [cncfq20adt['vertices'][z] for z in Z]
In [33]: from scipy.spatial import minkowski_distance
In [34]: lengthsXY = minkowski_distance(Xvert, Yvert); \
   ....: lengthsXZ = minkowski_distance(Xvert, Zvert); \
   ....: lengthsYZ = minkowski_distance(Yvert, Zvert)
```

We now create the weighted-adjacency matrix, which we store as a `lil_matrix`, and compute the shortest path between the requested vertices. We gather, in a list, all the vertices included in the computed path, and plot the resulting chain overlaid on the triangulation.

A word of warning:

The adjacency matrix we are about to compute is not the distance matrix. In the distance matrix A, we include, on each entry A[i, j], the distance between any vertex i to any vertex j, regardless of being connected by an edge or not. If this distance matrix is desired, the most reliable way to compute it is by means of the routine `distance_matrix` in the module `scipy.spatial`:

```
>>> from scipy.spatial import distance_matrix
>>> A = distance_matrix(cncfq20adt['vertices'],
                        cncfq20adt['vertices'])
```

```
In [35]: from scipy.sparse import lil_matrix; \
    ....: from scipy.sparse.csgraph import shortest_path
In [36]: nvert = len(cncfq20adt['vertices']); \
    ....: G = lil_matrix((nvert, nvert))
In [37]: for k in range(len(X)):
    ....:     G[X[k], Y[k]] = G[Y[k], X[k]] = lengthsXY[k]
    ....:     G[X[k], Z[k]] = G[Z[k], X[k]] = lengthsXZ[k]
    ....:     G[Y[k], Z[k]] = G[Z[k], Y[k]] = lengthsYZ[k]
    ....:
In [38]: dist_mat, pred = shortest_path(G, return_predecessors=True,
    ....:                               directed=True,
    ....:                               unweighted=False)
In [39]: index = 370; \
    ....: path = [370]
In [40]: while index != 197:
    ....:     index = pred[197, index]
    ....:     path.append(index)
    ....:
In [41]: print path
[380, 379, 549, 702, 551, 628, 467, 468, 469, 470, 632, 744, 764,
 799, 800, 791, 790, 789, 801, 732, 725, 570, 647, 177, 178, 179,
 180, 181, 182, 644, 571, 201, 200, 199, 197]
In [42]: ax = plt.subplot(111, aspect='equal'); \
    ....: tplot.plot(ax, **cncfq20adt)
In [43]: Xs = [cncfq20adt['vertices'][x][0] for x in path]; \
    ....: Ys = [cncfq20adt['vertices'][x][1] for x in path]
In [44]: ax.plot(Xs, Ys '-', linewidth=5, color='blue'); \
    ....: plt.show()
```

This gives the following diagram:

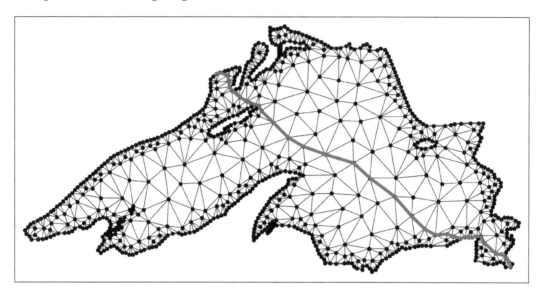

Geometric query problems

The fundamental problems in this category are as follows:

- Point location
- Nearest neighbor
- Range searching

Point location

The problem of point location is fundamental in computational geometry, given a partition of the space into disjoint regions, we need to query the region that contains a target location.

The most basic point location problems are those where the partition is given by a single geometric object—a circle or a polygon, for example. For those simple objects that have been constructed through any of the classes in the module sympy.geometry, we have two useful methods: .encloses_point and .encloses.

The former checks whether a point is interior to a source object (but not on the border), while the latter checks whether another target object has all its defining entities in the interior of the source object:

```
In [1]: from sympy.geometry import Point, Circle, Triangle
In [2]: P1 = Point(0, 0); \
   ...: P2 = Point(1, 0); \
   ...: P3 = Point(-1, 0); \
   ...: P4 = Point(0, 1)
In [3]: C = Circle(P2, P3, P4); \
   ...: T = Triangle(P1, P2, P3)
In [4]: print "Is P1 inside of C?", C.encloses_point(P1)
Is P1 inside of C? True
In [5]: print "Is T inside of C?", C.encloses(T)
Is T inside of C? False
```

Of special importance is this simple setting where the source object is a polygon. The routines in the sympy.geometry module get the job done, but at the cost of too many resources and too much time. A much faster way to approach this problem is by using the Path class from the libraries of matplotlib.path. Let's see how with a quick session. First, we create a representation of a polygon as a Path:

 For information about the class Path and its usage within the matplotlib libraries, refer to the official documentation at http://matplotlib.org/api/path_api. html#matplotlib.path.Path, as well as the tutorial at http://matplotlib.org/users/path_tutorial.html.

```
In [6]: import numpy as np, matplotlib.pyplot as plt; \
   ...: from matplotlib.path import Path; \
   ...: from chapter6 import read_poly; \
   ...: from scipy.spatial import ConvexHull
In [7]: superior = read_poly("superior.poly")
In [8]: hull = ConvexHull(superior['vertices'])
In [9]: my_polygon = Path([hull.points[x] for x in hull.vertices])
```

We can now ask whether a point (respectively, a sequence of points) is interior to the polygon. We accomplish this with either contains_point or contains_points:

```
In [10]: X = .25 * np.random.randn(100) + .5; \
   ....: Y = .25 * np.random.randn(100) - .5
In [11]: my_polygon.contains_points([[X[k], Y[k]] for k in
range(len(X))])
Out[11]:
```

```
array([False, False,  True, False,  True, False, False, False,  True,
       False, False, False,  True, False,  True, False, False, False,
        True, False,  True,  True, False, False, False, False, False,
       ...
        True, False,  True, False, False, False, False, False,  True,
        True, False,  True,  True,  True, False, False, False, False,
       False], dtype=bool)
```

More challenging point location problems arise when our space is partitioned by a complex structure. For instance, once a triangulation has been computed, and a random location is considered, we need to query for the triangle where our target location lies. In the module scipy.spatial, we have handy routines to perform this task over Delaunay triangulations created with scipy.spatial.Delaunay. In the following example, we track the triangles that contain a set of 100 random points in the domain:

```
In [12]: from scipy.spatial import Delaunay, tsearch
In [13]: tri = Delaunay(superior['vertices'])
In [14]: points = zip(X, Y)
In [15]: print tsearch(tri, points)
[ -1 687  -1 647  -1  -1  -1  -1 805 520 460 647 580  -1  -1  -1  -1
 304  -1  -1  -1  -1 108 723  -1  -1  -1  -1  -1  -1  -1 144 454  -1
  -1  -1 174 257  -1  -1  -1  -1  -1  52  -1  -1 985  -1 263  -1 647
  -1 314  -1  -1 104 144  -1  -1  -1  -1 348  -1 368  -1  -1  -1 988
  -1  -1  -1 348 614  -1  -1  -1  -1  -1  -1  -1 114  -1  -1 684  -1
 537 174 161 647 702 594 687 104  -1 144  -1  -1  -1 684  -1]
```

The same result is obtained with the method .find_simplex of the Delaunay object tri:

```
In [16]: print tri.find_simplex(points)
[ -1 687  -1 647  -1  -1  -1  -1 805 520 460 647 580  -1
  -1  -1  -1 304  -1  -1  -1  -1 108 723  -1  -1  -1  -1
  -1  -1  -1 144 454  -1  -1  -1 174 257  -1  -1  -1  -1
  -1  52  -1  -1 985  -1 263  -1 647  -1 314  -1  -1 104
 144  -1  -1  -1  -1 348  -1 368  -1  -1  -1 988  -1  -1
  -1 348 614  -1  -1  -1  -1  -1  -1  -1 114  -1  -1 684
  -1 537 174 161 647 702 594 687 104  -1 144  -1  -1  -1
 684  -1]
```

Note that, when a triangle is found, the routine reports its corresponding index in tri.simplices. If no triangle is found (which means the point is exterior to the convex hull of the triangulation), the index reported is -1.

Nearest neighbors

The problem of finding the Voronoi cell that contains a given location is equivalent to the search for the nearest neighbor in a set of seeds. We can always perform this search with a brute-force algorithm—and this is acceptable in some cases—but in general, there are more elegant and less complex approaches to this problem. The key lies in the concept of **k-d trees**—a special case of binary space partitioning structures for organizing points, conductive to fast searches.

In the SciPy stack, we have an implementation of k-d trees; the Python class `KDTree`, in the module `scipy.spatial`. This implementation is based on ideas published in 1999 by Maneewongvatana and Mount. It is initialized with the location of our input points. Once created, it can be manipulated and queried with the following methods and attributes:

- The methods are as follows:
    - `data`: This presents the input.
    - `leafsize`: This is the number of points at which the algorithm switches to brute-force. This value can be optionally offered in the initialization of the `KDTree` object.
    - `m`: This is the dimension of the space where the points are located.
    - `n`: This is the number of input points.
    - `maxes`: This indicates the highest values of each of the coordinates of the input points.
    - `mins`: This indicates the lowest values of each of the coordinates of the input points.

- The attributes are as follows:
    - `query(self, Q, p=2.0)`: This is the attribute that searches for the nearest neighbor, or a target location Q, using the structure of the k-d tree, with respect to the Minkowski p-distance.
    - `query_ball_point(self, Q, r, p=2.0)`: This is a more sophisticated query that outputs all points within the Minkowski p-distance r, from a target location Q.
    - `query_pairs(self, r, p=2.0)`: This finds all pairs of points whose Minkowski p-distance is, at most, r.
    - `query_ball_tree(self, other, r, p=2.0)`: This is similar to `query_pairs`, but it finds all pairs of points from two different k-d trees, which are at a Minkowski p-distance of, at least, r.

- ○   `sparse_distance_matrix(self, other, max_distance)`:
  This computes a distance matrix between two kd-trees, leaving
  as zero, any distance greater than `max_distance`. The output is
  stored in a sparse `dok_matrix`.

- ○   `count_neighbors(self, other, r, p=2.0)`: This attribute is an
  implementation of the two-point correlation designed by Gray and
  Moore, to count the number of pairs of points from two different k-d
  trees, which are at a Minkowski p-distance not larger than `r`. Unlike
  `query_ball`, this attribute does not produce the actual pairs.

There is a faster implementation of this object created as an extension type in Cython,
the `cdef` class `cKDTree`. The main difference is in the way the nodes are coded on
each case:

- For `KDTree`, the nodes are nested Python classes (the node being the top
  class, and both leafnode and innernode being subclasses that represent
  different kinds of nodes in the tree).

- For `cKDTree`, the nodes are C-type malloc'd structs, not classes. This makes
  the implementation much faster, at a price of having less control over a
  possible manipulation of the nodes.

Let's use this idea to solve a point location problem, and at the same time revisit the
Voronoi diagram from Lake Superior:

```
In [17]: from scipy.spatial import cKDTree, Voronoi, voronoi_plot_2d
In [18]: vor  = Voronoi(superior['vertices']); \
   ....: tree = cKDTree(superior['vertices'])
```

First, we query for the previous dataset, of 100 random locations, the seeds that are
closest to each of them:

```
In [19]: tree.query(points)
Out[19]:
(array([ 0.38942726,  0.05020313,  0.06987993,  0.2150344 ,
         0.16101652,  0.08485664,  0.33217896,  0.07993277,
         0.06298875,  0.07428273,  0.1817608 ,  0.04084714,
         0.0094284 ,  0.03073465,  0.01236209,  0.02395969,
         0.17561544,  0.16823951,  0.24555293,  0.01742335,
         0.03765772,  0.20490015,  0.00496507]),
 array([  3, 343, 311, 155, 370, 372, 144, 280, 197, 144, 251, 453,
         42, 233, 232, 371, 280, 311,   0, 307, 507,  49, 474, 370,
        114,   5,   1, 372, 285, 150, 361,  84,  43,  98, 418, 482,
        155, 144, 371, 113,  91,   3, 453,  91, 311, 412, 155, 156,
        251, 251,  22, 179, 394, 189,  49, 405, 453, 506, 407,  36,
        308,  33,  81,  46, 301, 144, 280, 409, 197, 407, 516]))
```

Note the output is a tuple with two `numpy.array`: the first one indicates the distances of each point to its closest seed (their nearest neighbors), and the second one indicates the index of the corresponding seed.

We can use this idea to represent the Voronoi diagram without a geometric description in terms of vertices, segments, and rays:

```
In [20]: X = np.linspace( 0.45,  0.50, 256); \
   ....: Y = np.linspace(-0.40, -0.35, 256); \
   ....: canvas = np.meshgrid(X, Y); \
   ....: points = np.c_[canvas[0].ravel(), canvas[1].ravel()]; \
   ....: queries = tree.query(points)[1].reshape(256, 256)
In [21]: ax1 = plt.subplot(121, aspect='equal'); \
   ....: voronoi_plot_2d(vor, ax=ax1); \
   ....: plt.xlim( 0.45,  0.50); \
   ....: plt.ylim(-0.40, -0.35)
Out[21]: (-0.4, -0.35)
In [22]: ax2 = plt.subplot(122, aspect='equal'); \
   ....: plt.gray(); \
   ....: plt.pcolor(X, Y, queries); \
   ....: plt.plot(vor.points[:,0], vor.points[:,1], 'ro'); \
   ....: plt.xlim( 0.45,  0.50); \
   ....: plt.ylim(-0.40, -0.35); \
   ....: plt.show()
```

This gives the following two different representations of the Voronoi diagram, in a small window somewhere along the North shore of Lake Superior:

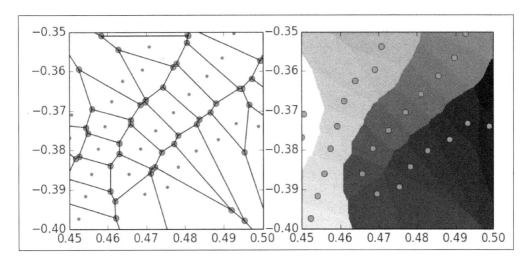

Range searching

A range searching problem tries to determine which objects of an input set intersect with a query object (that we call the **range**). For example, when given a set of points in the plane, which ones are contained inside a circle of radius r centered at a target location Q? We can solve this sample problem easily with the attribute `query_ball_point` from a suitable implementation of a k-d tree. We can go even further if the range is an object formed by the union of a sequence of different balls. The same attribute gets the job done, as the following code illustrates:

```
In [23]: points = np.random.rand(320, 2); \
    ....: range_points = np.random.rand(5, 2); \
    ....: range_radii = 0.1 * np.random.rand(5)
In [24]: tree = cKDTree(points); \
    ....: result = set()
In [25]: for k in range(5):
    ....:       point  = range_points[k]
    ....:       radius = range_radii[k]
    ....:       partial_query = tree.query_ball_point(point, radius)
    ....:       result = result.union(set(partial_query))
    ....:
In [26]: print result
set([130, 3, 166, 231, 40, 266, 2, 269, 120, 53, 24, 281, 26, 284])
In [27]: fig = plt.figure(); \
    ....: plt.axes().set_aspect('equal')
In [28]: for point in points:
    ....:       plt.plot(point[0], point[1], 'ko')
    ....:
In [29]: for k in range(5):
    ....:       point = range_points[k]
    ....:       radius = range_radii[k]
    ....:       circle = plt.Circle(point, radius, fill=False)
    ....:       fig.gca().add_artist(circle)
    ....:
In [30]: plt.show()
```

This gives the following diagram, where the small dots represent the locations of the search space, and the circles are the range. The query is the set of points located inside of the circles, computed by our algorithm:

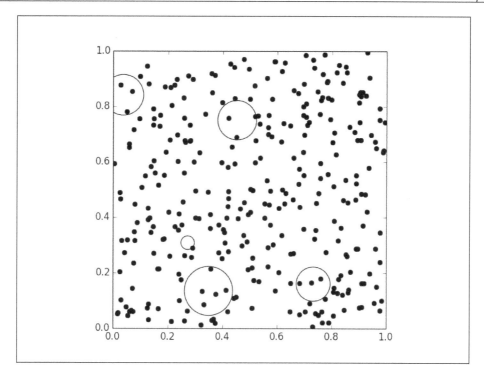

Problems in this setting vary from trivial to extremely complicated, depending on the input object types, range types, and query types. An excellent exposition of this subject is the survey paper *Geometric Range Searching and its Relatives*, published by Pankaj K. Agarwal and Jeff Erickson in 1999, by the American Mathematical Society Press, as part of the *Advances in Discrete and Computational Geometry*: proceedings of the 1996 AMS-IMS-SIAM joint summer research conference, Discrete and Computational geometry.

Dynamic problems

A dynamic problem is regarded as any of the problems in the previous two settings (static or query), but with the added challenge that objects are constantly being inserted or deleted. Besides solving the base problem, we need to take extra measures to assure that the implementation is efficient with respect to these changes.

To this effect, the implementations wrapped from the `Qhull` libraries in the module `scipy.spatial` are equipped to deal with the insertion of new points. We accomplish this by stating the option `incremental=True`, which basically suppresses the `qhull` control `'Qz'`, and prepares the output structure for these complex situations.

Let's illustrate this with a simple example. We start with the first ten vertices of Lake Superior, then we insert ten vertices at a time, and update the corresponding triangulation and Voronoi diagrams:

```
In [27]: from scipy.spatial import delaunay_plot_2d
In [28]: small_superior = superior['vertices'][:9]
In [29]: tri = Delaunay(small_superior, incremental=True); \
    ....: vor = Voronoi(small_superior, incremental=True)
In [30]: for k in range(4):
    ....:     tri.add_points(superior['vertices'][10*(k+1):10*(k+2)-1])
    ....:     vor.add_points(superior['vertices'][10*(k+1):10*(k+2)-1])
    ....:     ax1 = plt.subplot(4, 2, 2*k+1, aspect='equal')
    ....:     delaunay_plot_2d(tri, ax1)
    ....:     ax1.set_xlim( 0.00,  1.00)
    ....:     ax1.set_ylim(-0.70, -0.30)
    ....:     ax2 = plt.subplot(4, 2, 2*k+2, aspect='equal')
    ....:     voronoi_plot_2d(vor, ax2)
    ....:     ax2.set_xlim(0.0, 1.0)
    ....:     ax2.set_ylim(-0.70, -0.30)
    ....:
In [4]: plt.show()
```

This displays the following diagram:

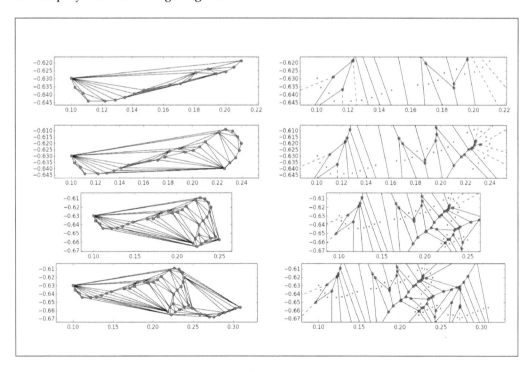

Numerical computational geometry

This field arose simultaneously among different groups of researchers seeking solutions to priori nonrelated problems. As it turns out, all the solutions they posed did actually have an important common denominator, they were obtained upon representing objects by means of parametric curves, parametric surfaces, or regions bounded by those. These scientists ended up unifying their techniques over the years, to finally define the field of **numerical computational geometry**. In this journey, the field received different names: **machine geometry**, **geometric modeling**, and the most widespread **computer aided geometric design** (**CAGD**).

It is used in computer vision, for example, for 3D reconstruction and movement outline. It is widely employed for the design and qualitative analysis of the bodies of automobiles, aircraft, or watercraft. There are many **computer-aided design** (**CAD**) software packages that facilitate interactive manipulation and solutions to many of the problems in this area. In this regard, any interaction with Python gets relegated to being part of the underlying computational engine behind the visualization or animation—which are none of the strengths of SciPy. For this reason, we will not cover visualization or animation applications in this book, and focus on the basic mathematics instead.

In that regard, the foundation of Numerical computational geometry is based on three key concepts: Bézier surfaces, Coons patches, and B-spline methods. In turn, the theory of Bézier curves plays a central role in the development of these concepts. They are the geometric standards for the representation of piecewise polynomial curves. In this section, we focus solely on the basic development of the theory of plane Bézier curves.

 The rest of the material is also beyond the scope of SciPy, and we therefore leave its exposition to more technical books. The best source in that sense is, without a doubt, the book *Curves and Surfaces for Computer Aided Geometric Design – A Practical Guide (5th ed.)*, by Gerald Farin, published by Academic Press under the Morgan Kauffman Series in Computer Graphics and Geometric Modeling.

Bézier curves

It all starts with the de Casteljau algorithm to construct parametric equations of an arc of a polynomial of order 3. In the submodule `matplotlib.path`, we have an implementation of this algorithm using the class `Path`, which we can use to generate our own user-defined routines to generate and plot plane Bézier curves:

```python
# file chapter6.py   ...continued
import matplotlib.pyplot as plt
import matplotlib.patches as patches
from matplotlib.path import Path
def bezier_parabola(P1, P2, P3):
    return Path([P1, P2, P3],
                [Path.MOVETO, Path.CURVE3, Path.CURVE3])
def bezier_cubic(P1, P2, P3, P4):
    return Path([P1, P2, P3, P4],
                [Path.MOVETO, Path.CURVE4,
                 Path.CURVE4, Path.CURVE4])
def plot_path(path, labels=None):
    Xs, Ys = zip(*path.vertices)
    fig = plt.figure()
    ax  = fig.add_subplot(111, aspect='equal')
    ax.set_xlim(min(Xs)-0.2, max(Xs)+0.2)
    ax.set_ylim(min(Ys)-0.2, max(Ys)+0.2)
    patch = patches.PathPatch(path, facecolor='none', linewidth=2)
    ax.add_patch(patch)
    ax.plot(Xs, Ys, 'o--', color='blue', linewidth=1)
    if labels:
        for k in range(len(labels)):
            ax.text(path.vertices[k][0]-0.1,
                path.vertices[k][1]-0.1,
                labels[k])
    plt.show()
```

Before we proceed, we need some explanation of the previous code:

- The de Casteljau algorithm for arcs of polynomials of order 2 is performed by creating a `Path` with the three control points as vertices, and the list `[Path.MOVETO, Path.CURVE3, Path.CURVE3]` as code. This ensures that the resulting curve starts at `P1` in the direction given by the segment `P1P2`, and ends at `P3` with direction given by the segment `P2P3`. If the three points are collinear, we obtain a segment containing them all. Otherwise, we obtain an arc of parabola.

- The de Casteljau algorithm for arcs of polynomials of order 3 is performed in a similar way to the previous case. We have four control points, and we create a Path with those as vertices. The code is the list [Path.MOVETO, Path.CURVE4, Path.CURVE4, Path.CURVE4], which ensures that the arc starts at P1 with direction given by the segment P1P2. It also ensures that the arc ends at P4 in the direction of the segment P3P4.

Let's test it with a few basic examples:

```
In [1]: import numpy as np; \
   ...: from chapter6 import *
In [2]: P1 = (0.0, 0.0); \
   ...: P2 = (1.0, 1.0); \
   ...: P3 = (2.0, 0.0); \
   ...: path_1 = bezier_parabola(P1, P2, P3); \
   ...: plot_path(path_1, labels=['P1', 'P2', 'P3'])
```

This gives the requested arc of parabola:

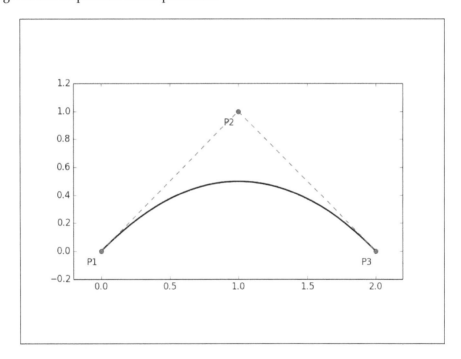

```
In [3]: P4 = (2.0, -1.0); \
   ...: P5 = (3.0, 0.0); \
   ...: path_2 = bezier_cubic(P1, P2, P4, P5); \
   ...: plot_path(path_2, labels=['P1', 'P2', 'P4', 'P5'])
```

This gives a nice arc of cubic as shown in the following diagram:

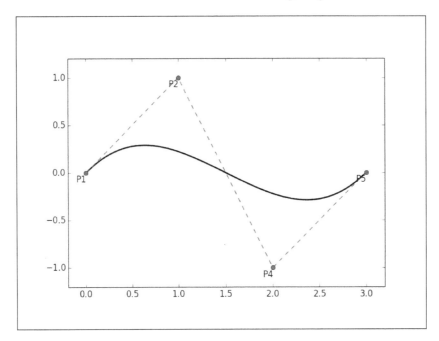

Higher degree curves are computationally expensive to evaluate. When complex paths are needed, we rather create them as a piecewise sequence of low order Bézier curves patched together — we call this object a Bézier spline. Notice that it is not hard to guarantee continuity on these splines. It is enough to make the end of each path the starting point of the next one. It is also easy to guarantee smoothness (at least up to the first derivative), by aligning the last two control points of one curve with the first two control points of the next one. Let's illustrate this with an example:

```
In [4]: Q1 = P5; \
   ...: Q2 = (4.0, 0.0); \
   ...: Q3 = (5.0, -1.0); \
   ...: Q4 = (6.0, 0.0); \
   ...: path_3 = bezier_cubic(P1, P2, P3, P5); \
   ...: path_4 = bezier_cubic(Q1, Q2, Q3, Q4); \
   ...: plot_path(Path.make_compound_path(path_3, path_4),
                  labels=['P1','P2','P3','P5=Q1',
                          'P5=Q1','Q2','Q3', 'Q4'])
```

This gives the following Bézier spline:

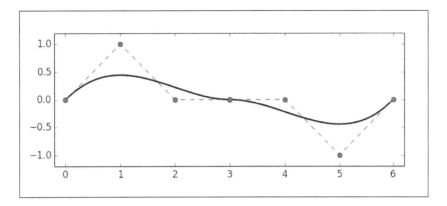

A clear advantage of representing curves as Bézier splines arises when we need to apply an affine transformation to a curve. For instance, if we required a counter-clockwise rotated version of the last curve computed, instead of performing the operation over all points of the curve, we simply apply the transformation to the control points and repeat the de Casteljau algorithm on the new controls:

```
In [5]: def rotation(point, angle):
   ...:         return (np.cos(angle)*point[0] - np.sin(angle)*point[1],
   ...:                 np.sin(angle)*point[0] + np.cos(angle)*point[1])
   ...:
In [6]: new_Ps = [rotation(P, np.pi/3) for P in path_3.vertices]; \
   ...: new_Qs = [rotation(Q, np.pi/3) for Q in path_4.vertices]; \
   ...: path_5 = bezier_cubic(*new_Ps); \
   ...: path_6 = bezier_cubic(*new_Qs); \
   ...: plot_path(Path.make_compound_path(path_5, path_6))
```

This displays the following result:

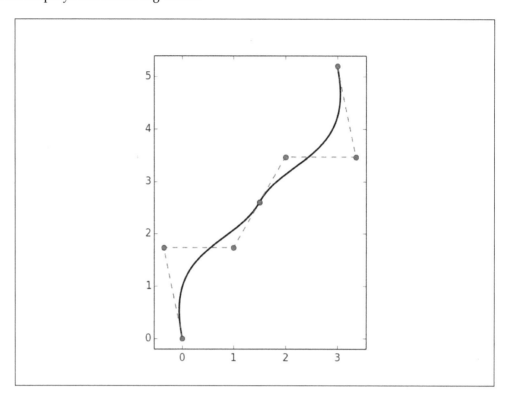

Summary

In this chapter, we have developed a brief incursion in the field of computational geometry, and we have mastered all the tools coded in the SciPy stack to effectively address the most common problems in this topic.

In the next two chapters, we will explore the capabilities of SciPy to work on applications of statistics, data mining, learning theory, and other techniques, to the field of quantitative data analysis.

Descriptive Statistics

7

This and the following chapter are mainly aimed at SAS, SPSS, or Minitab users, and especially those employing the languages R or S for statistical computing. We will develop an environment for working effectively in the field of data analysis, with the aid of IPython sessions powered up with the following resources from the SciPy stack:

- The probability and statistics submodule of the library of symbolic computations, `sympy.stats`.

- The two libraries of statistical functions `scipy.stats` and `scipy.stats.mstats` (the latter for data provided by masked arrays), together with the module `statsmodels`, for data exploration, estimation on statistical models, and performing statistical tests in a numerical setting. The package `statsmodels` uses, under the hood, the powerful library `patsy` to describe statistical models and building design matrices in Python (R or S users will find `patsy` compatible with their formula mini-language).

- For statistical inference, we again use `scipy.stats` and `statsmodels` (for frequentist and likelihood inference) and the module `pymc` that implements Bayesian statistical models and fitting algorithms, including Markov chain Monte Carlo.

- Two incredibly powerful libraries of high-level data manipulation tools.

    - The Python Data Analysis library `pandas`, created by Wes McKinney to address the useful functionalities of time series, data alignment, and the treatment of databases in a similar fashion to SQL.

    - The package `PyTables`, created by Francesc Alted, Ivan Vilata and others,
      for managing hierarchical datasets. It is designed to efficiently and easily cope with extremely large amounts of data.

- The clustering module `scipy.cluster` for vector quantization, the k-means algorithm, hierarchical and agglomerative clustering.

- A few SciPy toolkits (SciKits for short):

  ° `scikit-learn`: A set of Python modules for machine learning and data mining.

  ° `scikits.bootstrap`: Bootstrap confidence interval estimation routines.

An obvious knowledge of statistics is needed to follow the techniques in these chapters. Any good basic textbook with an excellent selection of examples and problems will suffice. For a more in-depth study of inference, we recommend the second edition of the book *Statistical Inference*, by George Casella and Roger L. Berger, published by Duxbury in 2002.

Documentation for the following `python` libraries can be obtained online through their corresponding official pages:

- `sympy.stats`: http://docs.sympy.org/dev/modules/stats.html.

- `scipy.stats` and `scipy.stats.mstats`: http://docs.scipy.org/doc/scipy/reference/stats.html for a list of functions and http://docs.scipy.org/doc/scipy/reference/tutorial/stats.html for a nice overview and tutorial.

- `scipy.cluster`: http://docs.scipy.org/doc/scipy/reference/cluster.html

- `PyTables`: http://www.pytables.org/

- `PyMC`: http://pymc-devs.github.io/pymc/index.html

The best way to get acquainted with Pandas is without a doubt the book *Python for Data Analysis: Data Wrangling with Pandas, NumPy, and IPython*, by Wes McKinney himself — the creator of this amazing library. Familiarity with SQL is a must and, for this, our recommendation is to get good training online.

One of the best resources to understand the topic of model estimation is the book *Methods of Statistical Model Estimation*, by Joseph Hilbe and Andrew Robinson. Although all the codes in this resource are written for R, they are easily portable to a combination of routines and classes from `scipy.stats`, `statsmodels`, `PyMC`, and `scikit-learn`.

The package `statsmodels` used to be part of the `scikit-learn` toolkit. Good documentation to learn the usage and power of this package, and its underlying library to describe statistical models (`patsy`) are always the tutorials offered by their creators online:

- `http://statsmodels.sourceforge.net/stable/`
- `http://patsy.readthedocs.org/en/latest/`

For the `scipy` toolkits, the best resource is found via their page at `http://scikits.appspot.com/`. Browsing the different toolkits will point us to good tutorials and further references. In particular, for `scikit-learn`, two must-reads are the official page at `http://scikit-learn.org/stable/`, and the seminal article *Scikit-learn: Machine Learning in Python*, by Fabian Pedregosa et al., published in the *Journal of Machine Learning Research* in 2011.

But as it is our custom in this book, we will develop the material from the point of view of the material itself. We have thus divided the exposition in two chapters, the first of which is concerned with the most basic topics in Probability and Statistics:

- **Probability**—Random variables and their distributions.
- **Data Exploration**.

In the next chapter, we will address more advanced topics in Statistics and Data Analysis:

- **Statistical inference**.
- **Machine learning**. The construction and study of systems that can learn from data. Machine learning focuses on prediction based on known properties learned from some training data.
- **Data mining**. Discovering patterns in large data sets. Data mining focuses on the discovery of priori unknown properties in the data.

Motivation

On Tuesday, September 8, 1857, the steamboat SS Central America left Havana at 9 A.M. for New York, carrying about 600 passengers and crew members. Inside this vessel, precious cargo was stored—a set of manuscripts by John James Audubon, and three tons of gold bars and coins. The manuscripts documented an expedition through the yet uncharted southwestern United States and California, and contained 200 sketches and paintings of its wildlife. The gold, fruit of many years of prospecting and mining during the California Gold Rush, was meant to start anew the lives of many of the passengers aboard.

On the 9th, the vessel ran into a storm which developed into a hurricane. The steamboat endured four hard days at sea, and by Saturday morning the ship was doomed. The captain arranged to have women and children taken off to the brig Marine, which offered them assistance at about noon. In spite of the efforts of the remaining crew and passengers to save the ship, the inevitable happened at about 8 P.M. that same day. The wreck claimed the lives of 425 men, and carried to the bottom of the sea the valuable cargo.

It was not until late 1980s that technology allowed recovery of shipwrecks in deep sea. However, no technology would be of any help without an accurate location of the site. In the following paragraphs, we would like to illustrate the power of the SciPy stack by performing a simple simulation. The objective is the creation of a dataset of possible locations for the wreck of the SS Central America. We mine this data to attempt to pinpoint the most probable target.

We simulate several possible paths of the steamboat (say 10,000 randomly generated possibilities), between 7 A.M. on Saturday, and 13 hours later, at 8 P.M on Sunday. At 7 A.M. on that Saturday, the ship's captain, William Herndon, took a celestial fix and verbally relayed the position to the schooner El Dorado. The fix was 31°25'N, 77°10'W. Since the ship was not operative at that point—no engine, no sails—for the next thirteen hours its course was solely subjected to the effect of ocean currents and winds. With enough information, it is possible to model the drift and leeway on different possible paths.

We start by creating a `DataFrame`—a computational structure that will hold all the values we need in a very efficient way. We do so with the help of the `pandas` libraries:

```
In [1]: from datetime import datetime, timedelta; \
   ...: from dateutil.parser import parse
In [2]: interval = [parse("9/12/1857 7 am")]
In [3]: for k in range(14*2-1):
   ...:        if k % 2 == 0:
   ...:            interval.append(interval[-1])
   ...:        else:
   ...:            interval.append(interval[-1] + timedelta(hours=1))
   ...:
In [4]: import numpy as np, pandas as pd
In [5]: herndon = pd.DataFrame(np.zeros((28, 10000)),
   ...:                        index = [interval, ['Lat', 'Lon']*14])
```

Each column of the DataFrame `herndon` is to hold the latitude and longitude of a possible path of the SS Central America, sampled every hour. For instance, to observe the first path, we issue the following command:

```
In [6]: herndon[0]
Out[6]:
1857-09-12 07:00:00   Lat    0
                      Lon    0
1857-09-12 08:00:00   Lat    0
                      Lon    0
1857-09-12 09:00:00   Lat    0
                      Lon    0
1857-09-12 10:00:00   Lat    0
                      Lon    0
1857-09-12 11:00:00   Lat    0
                      Lon    0
1857-09-12 12:00:00   Lat    0
                      Lon    0
1857-09-12 13:00:00   Lat    0
                      Lon    0
1857-09-12 14:00:00   Lat    0
                      Lon    0
1857-09-12 15:00:00   Lat    0
                      Lon    0
1857-09-12 16:00:00   Lat    0
                      Lon    0
1857-09-12 17:00:00   Lat    0
                      Lon    0
1857-09-12 18:00:00   Lat    0
                      Lon    0
1857-09-12 19:00:00   Lat    0
                      Lon    0
1857-09-12 20:00:00   Lat    0
                      Lon    0
Name: 0, dtype: float64
```

Let's populate this data following a similar analysis to that followed by the Columbus America Discovery Group, as explained by Lawrence D. Stone in the article *Revisiting the SS Central America Search*, from the 2010 International Conference on Information Fusion.

The celestial fix obtained by Capt. Herndon at 7 A.M. was taken with a sextant in the middle of a storm. There are some uncertainties in the estimation of latitude and longitude with this method and under those weather conditions, which are modeled by a bivariate normally distributed random variable with mean (0,0), and standard deviations of 0.9 nautical miles (for latitude) and 3.9 nautical miles (for longitude). We first create a random variable with those characteristics. Let's use this idea to populate the data frame with several random initial locations:

```
In [7]: from scipy.stats import multivariate_normal
In [8]: celestial_fix = multivariate_normal(cov = np.diag((0.9, 3.9)))
```

To estimate the corresponding celestial fixes, as well as all further geodetic computations, we will use the accurate formulas of Vincenty for ellipsoids, assuming a radius at the equator of `a = 6378137` meters and a flattening of the ellipsoid of `f = 1/298.257223563` (these figures are regarded as one of the standards for use in cartography, geodesy, and navigation, and are referred to by the community as the World Geodetic System WGS-84 ellipsoid).

A very good set of formulas coded in Python can be found at `https://github.com/blancosilva/Mastering-Scipy/blob/master/chapter7/Geodetic_py.py`. For a description and the theory behind these formulas, read the excellent survey on Wikipedia at `https://en.wikipedia.org/wiki/Vincenty%27s_formulae`.

In particular, for this example, we will be using Vincenty's direct formula that computes the resulting latitude `phi2`, longitude `L2`, and azimuth `s2`, of an object starting at latitude `phi1`, longitude `L1`, and traveling `s` meters with initial azimuth `s1`. Latitudes, longitudes, and azimuths are given in degrees, and distances in meters. We also use the convention of assigning negative values to the latitudes to the west. To apply the conversion from nautical miles or knots to their respective units in SI, we employ the system of units in `scipy.constants`.

```
In [9]: from Geodetic_py import vinc_pt; \
   ...: from scipy.constants import nautical_mile
In [10]: a = 6378137.0; \
   ....: f = 1./298.257223563
In [11]: for k in range(10000):
   ....:       lat_delta,lon_delta = celestial_fix.rvs() * nautical_mile
   ....:       azimuth = 90 - np.angle(lat_delta+1j*lon_delta, deg=True)
   ....:       distance = np.hypot(lat_delta, lon_delta)
```

```
....:       output = vinc_pt(f, a, 31+25./60,
....:                        -77-10./60, azimuth, distance)
....:       herndon.ix['1857-09-12 07:00:00',:][k] = output[0:2]
....:
In [12]: herndon.ix['1857-09-12 07:00:00',:]
Out[12]:
              0          1          2          3          4          5
Lat   31.455345  31.452572  31.439491  31.444000  31.462029  31.406287
Lon  -77.148860 -77.168941 -77.173416 -77.163484 -77.169911 -77.168462

              6          7          8          9        ...       9990
Lat   31.390807  31.420929  31.441248  31.367623      ...   31.405862
Lon  -77.178367 -77.187680 -77.176924 -77.172941      ...  -77.146794

           9991       9992       9993       9994       9995       9996
Lat   31.394365  31.428827  31.415392  31.443225  31.350158  31.392087
Lon  -77.179720 -77.182885 -77.159965 -77.186102 -77.183292 -77.168586

           9997       9998       9999
Lat   31.443154  31.438852  31.401723
Lon  -77.169504 -77.151137 -77.134298
[2 rows x 10000 columns]
```

We simulate the drift according to the formula `D = (V + leeway * W)`. In this formula, `V` (the ocean current) is modeled as a vector pointing about Northeast (around 45 degrees of azimuth) and a variable speed between 1 and 1.5 knots. The other random variable, `W`, represents the action of the winds in the area during the hurricane, which we choose to represent by directions ranging between south and east, and speeds with a mean of 0.2 knots and standard deviation of 1/30 knots. Both random variables are coded as bivariate normal. Finally, we have accounted for the leeway factor. According to a study performed on the blueprints of the SS Central America, we have estimated this leeway to be about 3 percent:

This choice of random variables to represent the ocean current
and wind differs from the ones used in the aforementioned paper.
In our version, we have not used the actual covariance matrices
as computed by Stone from data received from the Naval
Oceanographic Data Center. Rather, we have presented a very
simplified version.

```
In [13]: current = multivariate_normal((np.pi/4, 1.25),
   ....:                    cov=np.diag((np.pi/270, .25/3)))); \
   ....: wind = multivariate_normal((np.pi/4, .3),
   ....:                    cov=np.diag((np.pi/12, 1./30)))); \
   ....: leeway = 3./100
In [14]: for date in pd.date_range('1857-9-12 08:00:00',
   ....:                           periods=13, freq='1h'):
   ....:        before  = herndon.ix[date-timedelta(hours=1)]
   ....:        for k in range(10000):
   ....:            angle, speed = current.rvs()
   ....:            current_v = speed * nautical_mile * (np.cos(angle)
   ....:                         + 1j * np.sin(angle))
   ....:            angle, speed  = wind.rvs()
   ....:            wind_v = speed * nautical_mile * (np.cos(angle)
   ....:                         + 1j * np.sin(angle))
   ....:            drift = current_v + leeway * wind_v
   ....:            azimuth = 90 - np.angle(drift, deg=True)
   ....:            distance = abs(drift)
   ....:            output = vinc_pt(f, a, before.ix['Lat'][k],
   ....:                             before.ix['Lon'][k],
   ....:                             azimuth, distance)
   ....:            herndon.ix[date,:][k] = output[:2]
```

Let's plot the first three of those simulated paths:

```
In [15]: import matplotlib.pyplot as plt; \
   ....: from mpl_toolkits.basemap import Basemap
In [16]: m = Basemap(llcrnrlon=-77.4, llcrnrlat=31.2,urcrnrlon=-76.6,
   ....:             urcrnrlat=31.8, projection='lcc', lat_0 = 31.5,
   ....:             lon_0=-77, resolution='l', area_thresh=1000.)
```

```
In [17]: m.drawmeridians(np.arange(-77.4,-76.6,0.1),
    ....:                     labels=[0,0,1,1]); \
    ....: m.drawparallels(np.arange(31.2,32.8,0.1),labels=[1,1,0,0]);\
    ....: m.drawmapboundary()
In [18]: colors = ['r', 'b', 'k']; \
    ....: styles = ['-', '--', ':']
In [19]: for k in range(3):
    ....:     latitudes = herndon[k][:,'Lat'].values
    ....:     longitudes = herndon[k][:,'Lon'].values
    ....:     longitudes, latitudes = m(longitudes, latitudes)
    ....:     m.plot(longitudes, latitudes, color=colors[k],
    ....:         lw=3, ls=styles[k])
    ....:
In [20]: plt.show()
```

This presents us with these three possible paths followed by the SS Central America during its drift in the storm. As expected, they observe a north-easterly general direction, on occasion showing deviations from the effect of the strong winds:

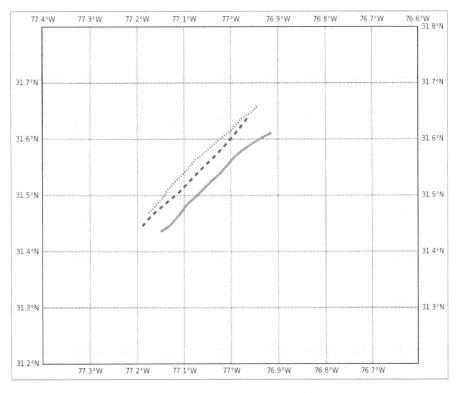

The focus of this simulation is, nonetheless, on the final location of all these paths. Let's plot them all on the same map first, for a quick visual evaluation:

```
In [21]: latitudes, longitudes = herndon.ix['1857-9-12 20:00:00'].values
In [22]: m = Basemap(llcrnrlon=-82., llcrnrlat=31, urcrnrlon=-76,
   ....:             urcrnrlat=32.5, projection='lcc', lat_0 = 31.5,
   ....:             lon_0=-78, resolution='h', area_thresh=1000.)
In [23]: X, Y = m(longitudes, latitudes)
In [24]: x, y = m(-81.2003759, 32.0405369)    # Savannah, GA
In [25]: m.plot(X, Y, 'ko', markersize=1); \
   ....: m.plot(x,y,'bo'); \
   ....: plt.text(x-10000, y+10000, 'Savannah, GA'); \
   ....: m.drawmeridians(np.arange(-82,-76,1), labels=[1,1,1,1]); \
   ....: m.drawparallels(np.arange(31,32.5,0.25), labels=[1,1,0,0]);\
   ....: m.drawcoastlines(); \
   ....: m.drawcountries(); \
   ....: m.fillcontinents(color='coral'); \
   ....: m.drawmapboundary(); \
   ....: plt.show()
```

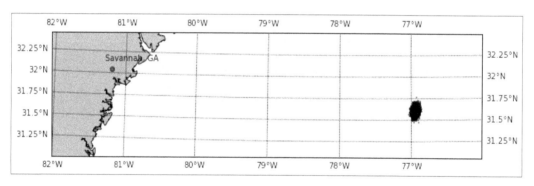

To obtain a better estimate of the true location of the shipwreck, it is possible to expand the simulation by using information from Captain Johnson of the Norwegian bark Ellen. This ship rescued several survivors at 8 A.M. on Sunday, at a recorded position of 31°55'N, 76°13'W. We can employ a similar technique to trace back to the location where the ship sunk using a reverse drift. For this case, the uncertainty in the celestial fix is modeled by a bivariate normal distribution with standard deviations of 0.9 (latitude) and 5.4 nautical miles (longitude).

A third simulation is also possible, using information from the El Dorado, but we do not factor this in our computations.

Since at this point the only relevant information is the location of the wreck, we do not need to keep the intermediate locations in our simulated paths. We record our data in a `pandas Series` instead:

```
In [26]: interval = []
In [27]: for k in range(10000):
   ....:         interval.append(k)
   ....:         interval.append(k)
   ....:
In [28]: ellen = pd.Series(index = [interval, ['Lat','Lon']*10000]);\
   ....: celestial_fix =multivariate_normal(cov=np.diag((0.9,5.4)));\
   ....: current = multivariate_normal((225, 1.25),
   ....:                                  cov=np.diag((2./3, .25/3)))
In [29]: for k in range(10000):
   ....:     lat_delta, lon_delta = celestial_fix.rvs()*nautical_mile
   ....:     azimuth = 90 - np.angle(lat_delta+1j*lon_delta, deg=True)
   ....:     distance = np.hypot(lat_delta, lon_delta)
   ....:     output = vinc_pt(f, a, 31+55./60,
   ....:                       -76-13./60, azimuth, distance)
   ....:     ellen[k] = output[0:2]
   ....:
In [30]: for date in pd.date_range('1857-9-13 07:00:00', periods=12,
   ....:                             freq='-1h'):
   ....:     for k in range(10000):
   ....:         angle, speed = current.rvs()
   ....:         output = vinc_pt(f, a, ellen[k,'Lat'],
   ....:                         ellen[k,'Lon'], 90-angle, speed)
   ....:         ellen[k] =output[0:2]
   ....:
```

The purpose of the simulation is the construction of a map that indicates the probability of finding the shipwreck depending on latitude and longitude. We can construct it by performing a kernel density estimation on the simulated data. The difficulty in this case lies in using the correct metric. Unfortunately, we are not able to create a metric based upon Vincenty's formulas in SciPy suitable for this operation, instead, we have two options:

- A linear approximation in a small area, using the routine `gaussian_kde` from the library `scipy.stats`

- A spherical approximation, using the class `KernelDensity` from the toolkit `scikit-learn`, imposing a Harvesine metric and a ball tree algorithm

The advantage of the first method is that it is faster, and the computations for optimal bandwidth are done internally. The second method is more accurate if we are able to provide the correct bandwidth. In any case, we prepare the data in the same way, using the simulation as training data:

```
In [31]: training_latitudes, training_longitudes = herndon.ix['1857-9-12
20:00:00'].values; \
   ....: training_latitudes = np.concatenate((training_latitudes,
   ....:                                   ellen[:,'Lat'])); \
   ....: training_longitudes = np.concatenate((training_longitudes,
   ....:                                   ellen[:,'Lon'])); \
   ....: values = np.vstack([training_latitudes,
   ....:                 training_longitudes]) * np.pi/180.
```

For the linear approximation, we perform the following computations:

```
In [32]: from scipy.stats import gaussian_kde
In [33]: kernel_scipy = gaussian_kde(values)
```

For the spherical approximation, and assuming a less than optimal bandwidth of 10^{-7}, we instead issue the following:

```
In [32]: from sklearn.neighbors import KernelDensity
In [33]: kernel_sklearn = KernelDensity(metric='haversine',
   ....:                                bandwidth=1.e-7,
   ....:                                kernel='gaussian',
   ....:                                algorithm='ball_tree')
   ....: kernel_sklearn.fit(values.T)
```

```
Out[33]:

KernelDensity(algorithm='ball_tree', atol=0, bandwidth=1e-07,
        breadth_first=True, kernel='gaussian', leaf_size=40,
        metric='haversine', metric_params=None, rtol=0)
```

From here all we need to do is generate a map, construct a grid on it, and using these values, project the corresponding evaluation of the computed kernel. This will give us a **probability density function (PDF)** of the corresponding distribution:

```
In [34]: plt.figure(); \
    ....: m = Basemap(llcrnrlon=-77.1, llcrnrlat=31.4,urcrnrlon=-75.9,
    ....:            urcrnrlat=32.6, projection='lcc', lat_0 = 32,
    ....:            lon_0=-76.5, resolution='l', area_thresh=1000);\
    ....: m.drawmeridians(np.arange(-77.5,-75.5,0.2),
    ....:                 labels=[0,0,1,1]); \
    ....: m.drawparallels(np.arange(31,33,0.2), labels=[1,1,0,0]); \
    ....: grid_lon, grid_lat = m.makegrid(25, 25); \
    ....: xy = np.vstack([grid_lat.ravel(),
    ....:                 grid_lon.ravel()]) * np.pi/180.
```

The computations of the PDF are done, depending of the kernel implemented, as follows:

```
In [35]: data = kernel_scipy(xy)
In [35]: data = np.exp(kernel_sklearn.score_samples(xy.T))
```

All that remains is to plot the results. We show the results of the first method, and leave the second as a nice exercise:

```
In [36]: levels = np.linspace(data.min(), data.max(), 6); \
    ....: data = data.reshape(grid_lon.shape)
In [37]: grid_lon, grid_lat = m(grid_lon, grid_lat); \
    ....: cs = m.contourf(grid_lon, grid_lat, data,
    ....:                 clevels=levels, cmap=plt.cm.Greys); \
    ....: cbar = m.colorbar(cs, location='bottom', pad="10%"); \
    ....: plt.show()
```

This presents us with a region of roughly 50 x 50 (nautical miles), colored by the corresponding density. The darker regions indicate a higher probability of finding the shipwreck:

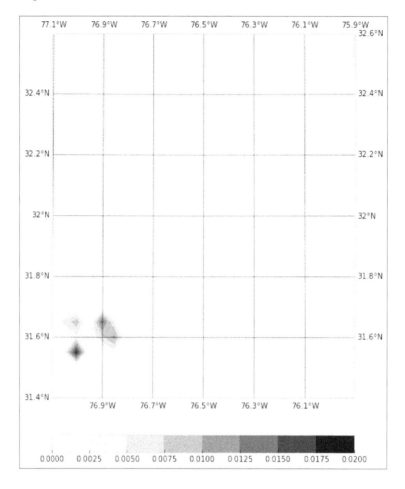

 The actual location of the remains of the SS Central America is at 31°35'N, 77°02'W, not too far from the results of our rough approximation — and as a matter of fact, very close to Captain Herndon's fix as communicated to the Marine.

This short motivational example illustrates the power of the SciPy stack to perform statistical simulations, store and manipulate the resulting data in optimal ways, and analyze them using state-of-the-art algorithms to extract valuable information. In the following pages, we will cover these techniques in more depth.

Probability

In the SciPy stack, we have two means for determining probability: a symbolic setting and a numerical setting. In this brief section, we are going to compare both with a sequence of examples.

For the symbolic treatment of random variables, we employ the module `sympy.stats`, while for the numerical treatment, we use the module `scipy.stats`. In both cases, the goal is the same—the instantiation of any random variable, and the following three kinds of operations on them:

- Description of the probability distribution of a random variable with numbers (parameters).
- Description of a random variable in terms of functions.
- Computation of associated probabilities.

Let's observe several situations through the scope of the two different settings.

Symbolic setting

Let's start with discrete random variables. For instance, let's consider several random variables used to describe the process of rolling three 6-sided dice, one 100-sided dice, and the possible outcomes:

```
In [1]: from sympy import var; \
   ...: from sympy.stats import Die, sample_iter, P, variance, \
   ...:                          std, E, moment, cdf, density, \
   ...:                          Exponential, skewness
In [2]: D6_1, D6_2, D6_3 = Die('D6_1', 6), Die('D6_2', 6), \
   ...:                     Die('D6_3', 6); \
   ...: D100 = Die('D100', 100); \
   ...: X = D6_1 + D6_2 + D6_3 + D100
```

We run a simulation, where we cast those four dice 20 times, and collect the sum of each throw:

```
In [3]: for item in sample_iter(X, numsamples=20):
   ...:         print item,
   ...:
45 50 84 43 44 84 102 38 90 94 35 78 67 54 20 64 62 107 59 84
```

Let's illustrate how easily we can compute probabilities associated with these variables. For instance, to calculate the probability that the sum of the three 6-sided dice amount to a smaller number than the throw of the 100-sided dice can be obtained as follows:

```
In [4]: P(D6_1 + D6_2 + D6_3 < D100)
Out[4]: 179/200
```

Conditional probabilities are also realizable, such as, "What is the probability of obtaining at least a 10 when throwing two 6-sided dice, if the first one shows a 5?":

```
In [5]: from sympy import Eq   # Don't use == with symbolic objects!
In [6]: P(D6_1 + D6_2 > 9, Eq(D6_1, 5))
Out[6]: 1/3
```

The computation of parameters of the associated probability distributions is also very simple. In the following session, we obtain the variance, standard deviation, and expected value of X, together with some other higher-order moments of this variable around zero:

```
In [7]: variance(X), std(X), E(X)
Out[7]: (842, sqrt(842), 61)
In [8]: for n in range(2,10):
   ...:     print "mu_{0} = {1}".format(n, moment(X, n, 0))
   ...:
mu_2 = 4563
mu_3 = 381067
mu_4 = 339378593/10
mu_5 = 6300603685/2
mu_6 = 1805931466069/6
mu_7 = 176259875749813/6
mu_8 = 29146927913035853/10
mu_9 = 586011570997109973/2
```

We can easily compute the probability mass function and cumulative density function too:

```
In [9]: cdf(X)            In [10]: density(X)
Out[9]:                   Out[10]:
{4: 1/21600,              {4: 1/21600,
 5: 1/4320,                5: 1/5400,
```

6: 1/1440,

7: 7/4320,

8: 7/2160,

9: 7/1200,

10: 23/2400,

11: 7/480,

12: 1/48,

13: 61/2160,

14: 791/21600,

15: 329/7200,

16: 1193/21600,

17: 281/4320,

18: 3/40,

. . .

102: 183/200,

103: 37/40,

104: 4039/4320,

105: 20407/21600,

106: 6871/7200,

107: 20809/21600,

108: 2099/2160,

109: 47/48,

110: 473/480,

111: 2377/2400,

112: 1193/1200,

113: 2153/2160,

114: 4313/4320,

115: 1439/1440,

116: 4319/4320,

117: 21599/21600,

118: 1}

6: 1/2160,

7: 1/1080,

8: 7/4320,

9: 7/2700,

10: 3/800,

11: 1/200,

12: 1/160,

13: 1/135,

14: 181/21600,

15: 49/5400,

16: 103/10800,

17: 53/5400,

18: 43/4320,

102: 1/100,

103: 1/100,

104: 43/4320,

105: 53/5400,

106: 103/10800,

107: 49/5400,

108: 181/21600,

109: 1/135,

110: 1/160,

111: 1/200,

112: 3/800,

113: 7/2700,

114: 7/4320,

115: 1/1080,

116: 1/2160,

117: 1/5400,

118: 1/21600}

Let's move onto continuous random variables. This short session computes the density and cumulative distribution function, as well as several parameters, of a generic exponential random variable:

```
In [11]: var('mu', positive=True); \
   ....: var('t'); \
   ....: X = Exponential('X', mu)
In [12]: density(X)(t)
Out[12]: mu*exp(-mu*t)
In [13]: cdf(X)(t)
Out[13]: Piecewise((1 - exp(-mu*t), t >= 0), (0, True))
In [14]: variance(X), skewness(X)
Out[14]: (mu**(-2), 2)
In [15]: [moment(X, n, 0) for n in range(1,10)]
Out[15]:
[1/mu,
 2/mu**2,
 6/mu**3,
 24/mu**4,
 120/mu**5,
 720/mu**6,
 5040/mu**7,
 40320/mu**8,
 362880/mu**9]
```

> For a complete description of the module sympy.stats with an exhaustive enumeration of all its implemented random variables, a good reference is the official documentation online at http://docs.sympy.org/dev/modules/stats.html.

Numerical setting

The description of a discrete random variable in the numerical setting is performed by the implementation of an object rv_discrete from the module scipy.stats. This object has the following methods:

- object.rvs to obtain samples
- object.pmf and object.logpmf to compute the probability mass function and its logarithm, respectively

- `object.cdf` and `object.logcdf` to compute the cumulative density function and its logarithm, respectively

- `object.sf` and `object.logsf` to compute the survival function (1-cdf) and its logarithm, respectively

- `object.ppf` and `object.isf` to compute the percent point function (the inverse of the CDF) and the inverse of the survival function

- `object.expect` and `object.moment` to compute expected value and other moments

- `object.entropy` to compute entropy

- `object.median`, `object.mean`, `object.var`, and `object.std` to compute the basic parameters (which can also be accessed with the method `object.stats`)

- `object.interval` to compute an interval with a given probability that contains a random realization of the distribution

We can then simulate an experiment with dice, similar to the previous section. In this setting, we represent dice by a uniform distribution on the set of the dice sides:

```
In [1]: import numpy as np, matplotlib.pyplot as plt; \
   ...: from scipy.stats import randint, gaussian_kde, rv_discrete
In [2]: D6 = randint(1, 7); \
   ...: D100 = randint(1, 101)
```

Symbolically, it was very simple to construct the sum of these four independent random variables. Numerically, we address the situation in a different way. Assume for a second that we do not know the kind of random variable we are to obtain. Our first step is usually to create a big sample—10,000 throws in this case, and produce a histogram with the results:

```
In [3]: samples = D6.rvs(10000) + D6.rvs(10000) \
   ...:           + D6.rvs(10000) + D100.rvs(10000)


In [4]: plt.hist(samples, bins=118-4); \
   ...: plt.xlabel('Sum of dice'); \
   ...: plt.ylabel('Frequency of each sum'); \
   ...: plt.show()
```

This gives the following screenshot that clearly indicates that our new random variable is not uniform:

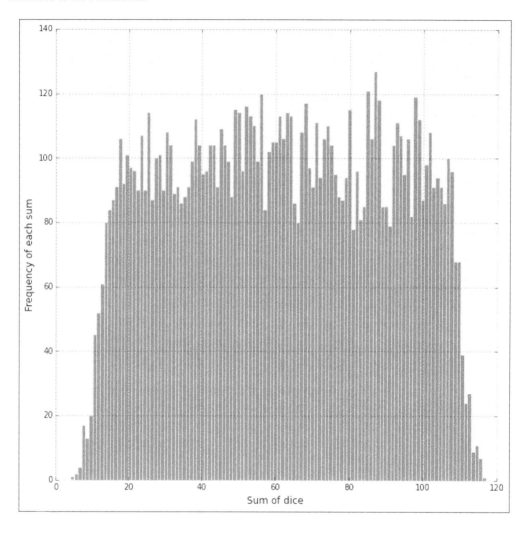

One way to approach this problem is to approximate the distribution of the variable from this data, and for that task, we use from the `scipy.stats` module the function `gaussian_kde`, which performs a kernel-density estimate using Gaussian kernels:

```
In [5]: kernel = gaussian_kde(samples)
```

This `gaussian_kde` object has methods similar to those of an actual random variable. To estimate the value of the corresponding probability of getting a 50, and the probability of obtaining a number greater than 100 in a throw of these four dice, we would issue, respectively:

```
In [6]: kernel(50)                    # The actual answer is 1/100
Out[6]: array([ 0.00970843])
In [7]: kernel.integrate_box_1d(0,100) # The actual answer is 177/200
Out[7]: 0.88395064140531865
```

Instead of estimating this sum of random variables, and again assuming we are not familiar with the actual result, we could create an actual random variable by defining its probability mass function in terms of the probability mass functions of the summands. The key? Convolution, of course, since the random variables for these dice are independent. The sample space is the set of numbers from 4 to 118 (`space_sum` in the following command), and the probabilities associated with each element (`probs_sum`) are computed as the convolution of the corresponding probabilities for each dice on their sample spaces:

```
In [8]: probs_6dice = D6.pmf(np.linspace(1,6,6)); \
   ...: probs_100dice = D100.pmf(np.linspace(1,100,100))
In [9]: probs_sum = np.convolve(np.convolve(probs_6dice,probs_6dice),
   ...:                     np.convolve(probs_6dice,probs_100dice)); \
   ...: space_sum = np.linspace(4, 118, 115)
In [10]: sum_of_dice = rv_discrete(name="sod",
    ...:                             values=(space_sum, probs_sum))
In [11]: sum_of_dice.pmf(50)
Out[11]: 0.0099999999999999985
In [12]: sum_of_dice.cdf(100)
Out[12]: 0.8950000000000000057
```

Data exploration

Data exploration is generally performed by presenting a meaningful synthesis of its distribution—it could be through a sequence of graphs, by describing it with a set of numerical parameters, or by approximating it with simple functions. Now let's explore different possibilities, and how to accomplish them with different tools in the SciPy stack.

Picturing distributions with graphs

The type of graph depends on the type of variable (categorical, quantitative, or dates).

Bar plots and pie charts

When our data is described in terms of categorical variables, we often use pie charts or bar graphs to represent it. For example, we access the Consumer Complaint Database from the Consumer Financial Protection Bureau, at http://catalog. data.gov/dataset/consumer-complaint-database. The database was created in February 2014 to contain complaints received by the Bureau about financial products and services. In its updated version in March of the same year, it consisted of almost 300,000 complaints acquired since November 2011:

```
In [1]: import numpy as np, pandas as pd, matplotlib.pyplot as plt
In [2]: data = pd.read_csv("Consumer_Complaints.csv",
   ...:                    low_memory=False, parse_dates=[8,9])
In [3]: data.head()
Out[3]:
   Complaint ID                  Product  \
0       1015754          Debt collection
1       1015827          Debt collection
2       1016131          Debt collection
3       1015974  Bank account or service
4       1015831  Bank account or service


                     Sub-product  \
0  Other (phone, health club, etc.)
1                              NaN
2                          Medical
3                 Checking account
4                 Checking account


                              Issue  \
0  Cont'd attempts collect debt not owed
1      Improper contact or sharing of info
2        Disclosure verification of debt
3  Problems caused by my funds being low
```

```
4    Problems caused by my funds being low


                                   Sub-issue State   ZIP code  \
0                           Debt was paid    NY      11433
1        Contacted me after I asked not to   VT       5446
2    Right to dispute notice not received   TX      77511
3                                     NaN    FL      32162
4                                     NaN    TX      77584


    Submitted via Date received Date sent to company  \
0            Web    2014-09-05          2014-09-05
1            Web    2014-09-05          2014-09-05
2            Web    2014-09-05          2014-09-05
3            Web    2014-09-05          2014-09-05
4            Web    2014-09-05          2014-09-05


                                        Company  \
0                     Enhanced Recovery Company, LLC
1                     Southwest Credit Systems, L.P.
2                     Expert Global Solutions, Inc.
3    FNIS (Fidelity National Information Services, ...
4                                     JPMorgan Chase


                  Company response Timely response?  \
0    Closed with non-monetary relief          Yes
1                     In progress             Yes
2                     In progress             Yes
3          Closed with explanation            Yes
4          Closed with explanation            Yes


    Consumer disputed?
0              NaN
1              NaN
2              NaN
3              NaN
4              NaN
```

We downloaded the database in the simplest format they offer, a comma-separated-value file. We do so from `pandas` with the command `read_csv`. If we want to download the database in other formats (JSON, excel, and so on), we only need to adjust the reading command accordingly:

```
>>> pandas.read_csv("Consumer_Complaints.csv")    # CSV
>>> pandas.read_json("Consumer_Complaints.json")  # JSON
>>> pandas.read_excel("Consumer_Complaints.xls")  # XLS
```

Even more amazingly so, it is possible to retrieve the data online (no need to save it to our computer), if we know its URL:

```
>>> url1 = "https://data.consumerfinance.gov/api/views"
>>> url2 = "/x94z-ydhh/rows.csv?accessType=DOWNLOAD"
>>> url = url1 + url2
>>> data = pd.read_csv(url)
```

If the database contains trivial data, the parser might get confused with the corresponding `dtype`. In that case, we request the parser to try and resolve that situation, at the expense of using more memory resources. We do so by including the optional Boolean flag `low_memory=False`, as was the case in our running example.

Also, note how we specified `parse_dates=True`. An initial exploration of the file with the data showed that both the eighth and ninth columns represent dates. The library `pandas` has great capability to manipulate those without resorting to complicated `str` operations, and thus we indicate to the reader to transform those columns to the proper format. This will ease our treatment of the data later on.

Now let's present a bar plot indicating how many of these different complaints per company are on each `Product`:

```
In [4]: data.groupby('Product').size()
Out[4]:
Product
Bank account or service    35442
Consumer loan               8187
Credit card                39114
Credit reporting           35761
Debt collection            37737
Money transfers             1341
Mortgage                  118037
```

```
Payday loan                   1228
Student loan                  8659
dtype: int64
In [5]: _.plot(kind='barh'); \
   ...: plt.tight_layout(); \
   ...: plt.show()
```

This gives us the following interesting horizontal bar plot, showing the volume of complaints for each different product, from November 2011 to September 2014:

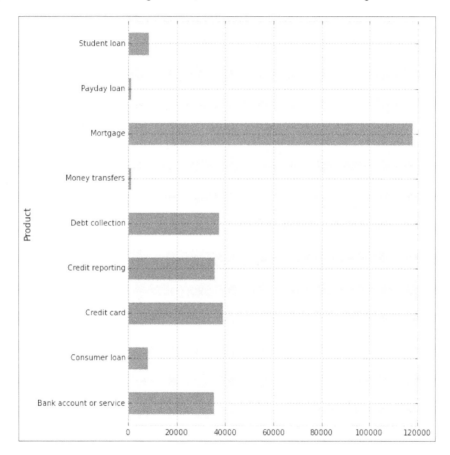

The groupby method on pandas dataframes is equivalent to GROUP BY in SQL. For a complete explanation of all SQL commands and their equivalent dataframe methods in pandas, there is a great resource online at http://pandas.pydata.org/pandas-docs/stable/comparison_with_sql.html.

Another informative bar plot is achieved by stacking the bars properly. For instance, if we focus on complaints about mortgages in the Midwest during the years 2012 and 2013, we could issue the following commands:

```
In [6]: midwest = ['ND', 'SD', 'NE', 'KS', 'MN', 'IA', \
                   'MO', 'IL', 'IN', 'OH', 'WI', 'MI']
In [7]: df = data[data.Product == 'Mortgage']; \
   ...: df['Year'] = df['Date received'].map(lambda t: t.year); \
   ...: df = df.groupby(['State','Year']).size(); \
   ...: df[midwest].unstack().ix[:, 2012:2013]
Out[7]:
Year    2012   2013
State
ND        14     20
SD        33     34
NE       109    120
KS       146    169
MN       478    626
IA        99    125
MO       519    627
IL      1176   1609
IN       306    412
OH      1047   1354
WI       418    523
MI      1457   1774
In [8]: _.plot(kind="bar", stacked=True, colormap='Greys'); \
   ...: plt.show()
```

The graph clearly illustrates how the year 2013 gave rise to a much higher volume of complaints in these states:

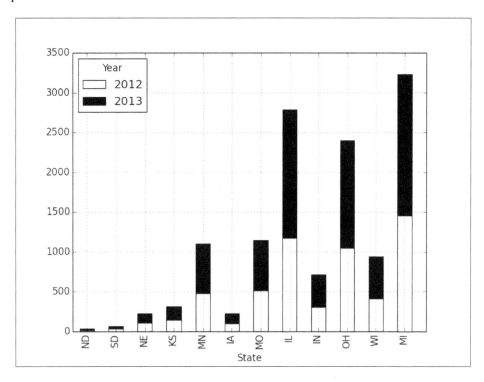

We may be inclined to think, in view of the previous graphs, that in every state or territory of the United States, mortgages are the number one complaint. A pie chart showing the volumes of complaints per product in Puerto Rico alone, from November 2011 until September 2014, tells us differently:

```
In [9]: data[data.State=='PR'].groupby('Product').size()
Out[9]:
Product
Bank account or service      81
Consumer loan                20
Credit card                 149
Credit reporting            139
Debt collection              62
Mortgage                    110
Student loan                 11
```

```
Name: Company, dtype: int64
In [10]: _.plot(kind='pie', shadow=True, autopct="%1.1f%%"); \
    ....: plt.axis('equal'); \
    ....: plt.tight_layout(); \
    ....: plt.show()
```

The diagram illustrates how credit cards and credit reports are the main source for complaints on these islands:

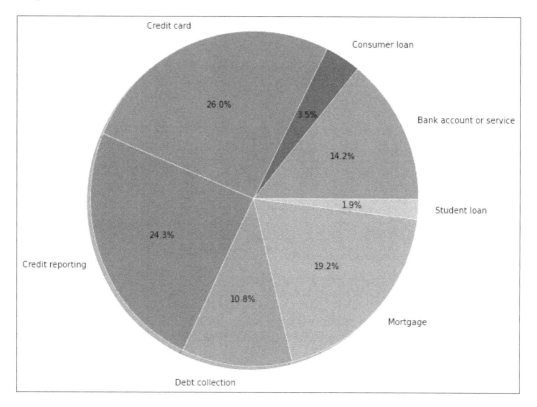

Histograms

For quantitative variables, we employ a histogram. In the previous section, we saw an example of the construction of a histogram from 10,000 throws of four dice. In this section, we produce another histogram from within `pandas`. In this case, we would like to present a histogram that analyzes the ratio of daily complaints about credit cards against the daily complaints on mortgages:

```
In [11]: df = data.groupby(['Date received', 'Product']).size(); \
    ....: df = df.unstack()
```

```
In [12]: ratios = df['Mortgage'] / df['Credit card']
In [13]: ratios.hist(bins=50); \
    ....: plt.show()
```

The resulting graph indicates, for instance, that there are a few days in which the number of complaints on mortgages is about 12 times the number of complaints on credit cards. It also shows that the most frequent situation is that of days in which the number of complaints on mortgages roughly triplicates the number of credit card complaints:

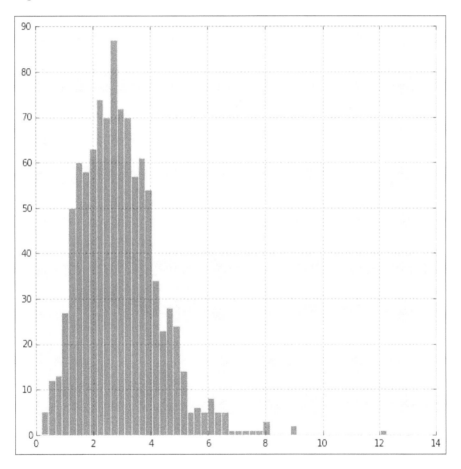

Time plots

For variables measured at intervals over time, we employ a time plot. The library `pandas` handles these beautifully. For instance, to observe the amount of daily complaints received from January 1, 2012 to December 31, 2013, we issue the following command:

```
In [14]: ts = data.groupby('Date received').size(); \
    ....: ts = ts['2012':'2013']; \
    ....: ts.plot(); \
    ....: plt.ylabel('Daily complaints'); \
    ....: plt.show()
```

Note both the oscillating nature of the graph, as well as the slight upward trend in complaints during this period. We also observe what looks like a few outliers—one in March 2012, a few more in between May and June of the same year, and a few more in January and February 2013:

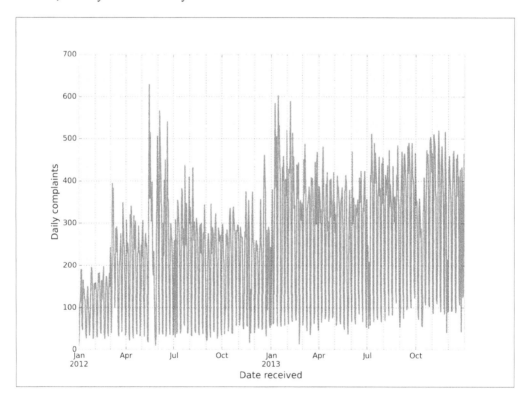

Describing distributions with numbers and boxplots

We request the usual parameters for each dataset:

- Mean (arithmetic, geometric or harmonic) and median to measure the center of the data
- Quartiles, variance, standard deviation, or standard error of the mean, to measure the spread of the data
- Central moments, skewness, and kurtosis to measure the degree of symmetry in the distribution of the data
- Mode to find the most common values in the data
- Trimmed versions of the previous parameters, to better describe the distribution of the data reducing the effect of outliers

A good way to present some of the preceding information is by means of the five-number summary, or with a boxplot.

Let's illustrate how to achieve these basic measurements both in `pandas` (left columns) and with the `scipy.stats` libraries (right columns):

```
In [15]: ts.describe()          In [16]: import scipy.stats
Out[15]:
count    731.000000             In [17]: scipy.stats.describe(ts)
mean     247.333789             Out[17]:
std      147.02033              (731,
min        9.000000             (9, 628),
25%      101.000000              247.33378932968537,
50%      267.000000              21614.97884301858,
75%      364.000000              0.012578861206579875,
max      628.000000              -1.1717653990872499)
dtype: float64
```

This second output presents us with count (`731 = 366 + 365`), minimum and maximum values (`min=9`, `max=628`), arithmetic mean (`247`), unbiased variance (`21614`), biased skewness (`0.0126`), and biased kurtosis (`-1.1718`).

Other computations of parameters, with both `pandas` (mode and standard deviation) and `scipy.stats` (standard error of the mean and trimmed variance of all values between 50 and 600):

```
In [18]: ts.mode()            In [20]: scipy.stats.sem(ts)
Out[18]:                      Out[20]: 5.4377435122312807
0    59
dtype: int64                  In [21]: scipy.stats.tvar(ts, [50, 600])
                              Out[21]: 17602.318623850999
In [19]: ts.std()
Out[19]: 147.02033479426774
```

For a complete description of all statistical functions in the `scipy.stats` library, the best reference is the official documentation at `http://docs.scipy.org/doc/scipy/reference/stats.html`.

It is possible to ignore NaN values in the computations of parameters. Most dataframe and series methods in pandas do that automatically. If a different behavior is required, we have the ability to substitute those NaN values for anything we deem appropriate. For instance, if we wanted any of the previous computations to take into account all dates and not only the ones registered, we could impose a value of zero complaints in those events. We do so with the method `dataframe.fillna(0)`.

With the library `scipy.stats`, if we want to ignore NaN values in an array, we use the same routines appending the keyword nan before their name:

```
>>> scipy.stats.nanmedian(ts)
267.0
```

In any case, the time series we computed shows absolutely no NaNs—there were at least nine daily financial complaints each single day in the years 2012 and 2013.

Going back to complaints about mortgages on the Midwest, to illustrate the power of boxplots, we are going to inquire in to the number of monthly complaints about mortgages in the year 2013, in each of those states:

```
In [22]: in_midwest = data.State.map(lambda t: t in midwest); \
    ....: mortgages = data.Product == 'Mortgage'; \
    ....: in_2013 =data['Date received'].map(lambda t: t.year==2013);\
    ....: df = data[mortgages & in_2013 & in_midwest]; \
    ....: df['month'] = df['Date received'].map(lambda t: t.month); \
    ....: df = df.groupby(['month', 'State']).size(); \
    ....: df.unstack()
Out[22]:
```

State	IA	IL	IN	KS	MI	MN	MO	ND	NE	OH	SD	WI
month												
1	11	220	40	12	183	99	91	3	13	163	5	58
2	14	160	37	16	180	45	47	2	12	120	NaN	37
3	7	138	43	18	184	52	57	3	11	131	5	50
4	14	148	33	19	185	55	52	2	14	104	3	48
5	14	128	44	16	172	63	57	2	8	109	3	43
6	20	136	47	13	164	51	47	NaN	13	116	7	52
7	5	127	30	16	130	57	62	2	11	127	5	39
8	11	133	32	15	155	64	55	NaN	8	120	1	51
9	10	121	24	16	99	31	55	NaN	8	109	NaN	37
10	9	96	35	12	119	50	37	3	10	83	NaN	35
11	4	104	22	10	96	22	39	2	6	82	3	32
12	6	98	25	6	107	37	28	1	6	90	2	41

```
In [23]: _.boxplot(); \
    ....: plt.show()
```

This boxplot illustrates how, among the states in the Midwest—Illinois, Ohio, and Michigan have the largest amount of monthly complaints on mortgages. In the case of Michigan (MI), for example, the corresponding boxplot indicates that the spread goes from 96 to 185 monthly complaints. The median number of monthly complaints in that state is about 160. The first and third quartiles are, respectively, 116 and 180:

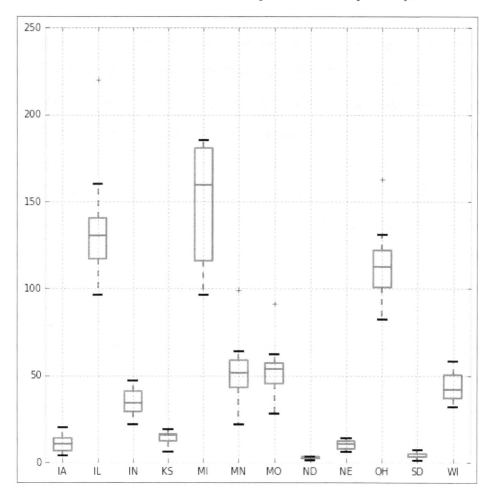

A violin plot is a boxplot with a rotated kernel density estimation on each side. This shows the probability density of the data at different values. We can obtain these plots with the graphical routine `violinplot` from the `statsmodels` submodule `graphics.boxplots`. Let's illustrate this kind of plot with the same data as before:

Another option is a combination of a violin plot with a line-scatter plot of all individual data points. We call this a **bean plot**, and we have an implementation with the routine `beanplot` in the same submodule `statsmodels.graphics.boxplot`.

```
In [24]: from statsmodels.graphics.boxplots import violinplot
In [25]: df = df.unstack().fillna(0)
In [26]: violinplot(df.values, labels=df.columns); \
   ....: plt.show()
```

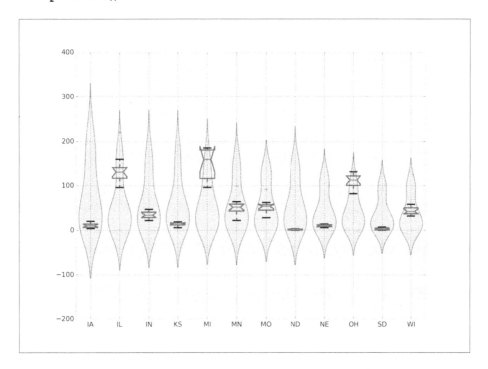

Relationship between quantitative variables

To express the relationship between two quantitative variables, we resort to three techniques:

- A **scatterplot** to visually identify that relationship
- The computation of a **correlation coefficient** that expresses how likely that relationship is to be formulated by a linear function
- A **regression** function as a means to predict the value of one of the variables with respect to the other

Scatterplots and correlation

For example, we are going to try to find any relation among the number of complaints on mortgages among four populous states—Illinois, New York, Texas, and California—and the territory of Puerto Rico. We will compare the number of complaints in each month from December 2011 to September 2014:

```
In [27]: from pandas.tools.plotting import scatter_matrix
In [28]: def year_month(t):
    ....:     return t.tsrftime("%Y%m")
    ....:
In [29]: states = ['PR', 'IL', 'NY', 'TX', 'CA']; \
    ....: states = data.State.map(lambda t: t in states); \
    ....: df = data[states & mortgages]; \
    ....: df['month'] = df['Date received'].map(year_month); \
    ....: df.groupby(['month', 'State']).size().unstack()
Out[29]:
```

State	CA	IL	NY	PR	TX
month					
2011/12	288	34	90	7	63
2012/01	444	77	90	2	104
2012/02	446	80	110	3	115
2012/03	605	78	179	3	128
2012/04	527	69	188	5	152
2012/05	782	100	242	3	151
2012/06	700	107	204	NaN	153
2012/07	668	114	198	3	153
2012/08	764	108	228	3	187
2012/09	599	92	192	1	140
2012/10	635	125	188	2	150
2012/11	599	99	145	6	130
2012/12	640	127	219	2	128
2013/01	1126	220	342	3	267
2013/02	928	160	256	4	210
2013/03	872	138	270	1	181
2013/04	865	148	254	5	200
2013/05	820	128	242	4	198
2013/06	748	136	232	1	237
2013/07	824	127	258	5	193

```
2013/08    742   133   236     3   183
2013/09    578   121   203   NaN   178
2013/10    533    96   193     2   123
2013/11    517   104   173     1   141
2013/12    463    98   163     4   152
2014/01    580    80   201     3   207
2014/02    670   151   189     4   189
2014/03    704   141   245     4   182
2014/04    724   146   271     4   212
2014/05    559   110   212    10   175
2014/06    480   107   226     6   141
2014/07    634   113   237     1   171
2014/08    408   105   166     5   118
2014/09      1   NaN     1   NaN     1
In [30]: df = _.dropna(); \
   ....: scatter_matrix(df); \
   ....: plt.show()
```

This gives the following matrix of scatter plots between the data of each pair of states, and the histogram of the same data for each state:

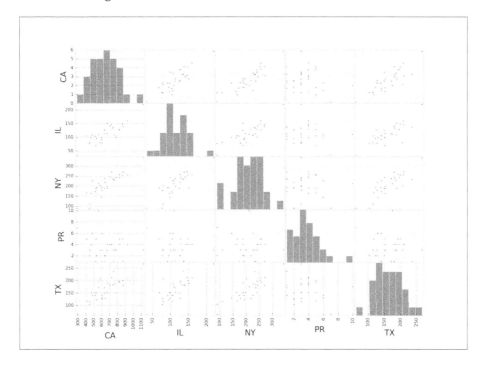

For a grid of scatter plots with confidence ellipses added, we can use the routine `scatter_ellipse` from the graphics module `graphics.plot_grids` of the package `statsmodels`:

```
In [31]: from statsmodels.graphics.plot_grids import scatter_ellipse
In [32]: scatter_ellipse(df, varnames=df.columns); \
   ....: plt.show()
```

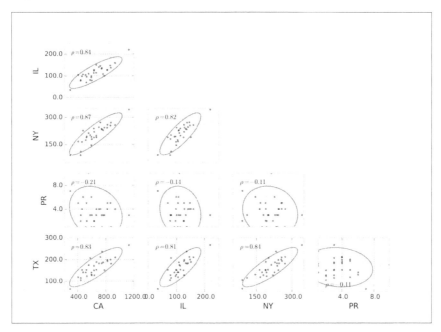

Note how each image comes with an extra piece of information. This is the corresponding correlation coefficient of the two variables (Pearson's, in this case). Data that appears to be almost perfectly aligned gets a correlation coefficient very close to 1 in the absolute value. We can obtain all these coefficients as the `pandas` dataframe method `corr` as well:

```
In [33]: df.corr(method="pearson")
Out[33]:
State          CA         IL         NY         PR         TX
State
CA       1.000000   0.844015   0.874480  -0.210216   0.831462
IL       0.844015   1.000000   0.818283  -0.141212   0.805006
NY       0.874480   0.818283   1.000000  -0.114270   0.837508
PR      -0.210216  -0.141212  -0.114270   1.000000  -0.107182
TX       0.831462   0.805006   0.837508  -0.107182   1.000000
```

Besides the standard Pearson correlation coefficients, this method allows us to compute Kendall Tau for ordinal data (`kendall`) or Spearman rank-order (`spearman`):

In the module `scipy.stats`, we also have routines for the computation of these correlation coefficients.

- `pearsonr` for Pearson's correlation coefficient and the p-value for testing noncorrelation
- `spearmanr` for the Spearman rank-order correlation coefficient and the p-value to test for noncorrelation
- `kendalltau` for Kendall's tau
- `pointbiserial` for the point biserial correlation coefficient and the associated p-value

Another possibility of visually displaying the correlation is by means of color grids. The graphic routine `plot_corr` from the submodule `statsmodels.graphics.correlation` gets the job done:

```
In [34]: from statsmodels.graphics.correlation import plot_corr
In [35]: plot_corr(df.corr(method='spearman'),
   ....:           xnames=df.columns.tolist()); \
   ....: plt.show()
```

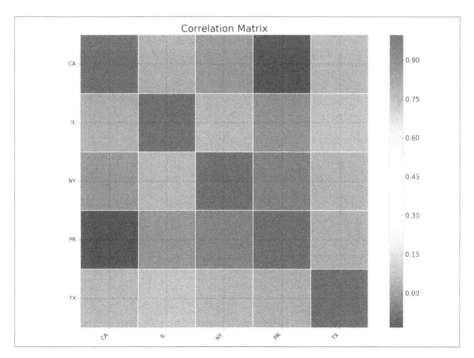

The largest correlation happens between the states of New York and California (0.874480). We will use the data corresponding to these two states for our subsequent examples in the next section.

Regression

Scatter plots helped us identify situations where the data could potentially be related by a functional relationship. This allows us to formulate a rule to predict the value of a variable knowing the other. When we suspect that such a formula exists, we want to find a good approximation to it.

We follow in this chapter the jargon of statisticians and data analysts, so rather than referring to this as an **approximation**, we will call it a **regression**. We also append an adjective indicating the kind of formula we seek. That way, we refer to linear regression if the function is linear, polynomial regression if it is a polynomial, and so on. Also, regressions do not necessarily involve only one variable in terms of another single variable. We thus differentiate between single-variable regression and multiple regression. Let's explore different settings for regression, and how to address them from the SciPy stack.

Ordinary linear regression for moderate-sized datasets

In any given case, we can employ the tools we learned during our exploration of approximation and interpolation in the least-squares sense, in *Chapters 1*, *Numerical Linear Algebra*, and *Chapter 2*, *Interpolation and Approximation*. There are many tools in the two libraries `scipy.stack` and `statsmodels`, as well as in the toolkit `scikit-learn` to perform this operation and associated analysis:

- A basic routine to compute ordinary least-square regression lines, `linregress`, in the `scipy.stats` library.

- The class `LinearRegression` from the `scikit-learn` toolkit, at `sklearn.linear_model`.

- A set of different regression routines in the `statsmodel` libraries, with the assistance of the `patsy` package.

Let's start with the simplest method via `linregress` in the `scipy.stats` library. We want to explore the almost-linear relationship between the number of monthly complaints on mortgages in the states of California and New York:

```
In [36]: x, y = df[['NY', 'CA']].T.values
In [37]: slope,intercept,r,p,std_err = scipy.stats.linregress(x,y); \
   ....: print "Formula: CA = {0} + {1}*NY".format(intercept, slope)
Formula: CA = 65.7706648926 + 2.82130682025*NY
```

```
In [38]: df[['NY', 'CA']].plot(kind='scatter', x='NY', y='CA'); \
    ....: xspan = np.linspace(x.min(), x.max()); \
    ....: plt.plot(xspan, intercept + slope * xspan, 'r-', lw=2); \
    ....: plt.show()
```

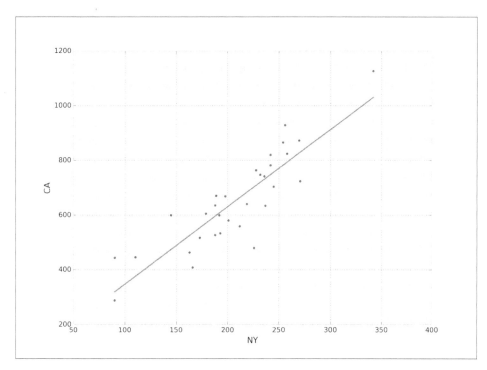

This is exactly the same result we obtain by using `LinearRegression` from the
`scikit-learn` toolkit:

```
In [39]: from sklearn.linear_model import LinearRegression
In [40]: model = LinearRegression()
In [41]: x = np.resize(x, (x.size, 1))
In [42]: model.fit(x, y)
Out[42]: LinearRegression(copy_X=True, fit_intercept=True,
    ....:                     normalize=False)
In [43]: model.intercept_
Out[43]: 65.770664892647233
In [44]: model.coef_
Out[44]: array([ 2.82130682])
```

For a more advanced treatment of this ordinary least-square regression line, offering more informative plots and summaries, we use the routine `ols` from `statsmodels`, and some of its awesome plotting utilities:

```
In [45]: import statsmodels.api as sm; \
   ....: from statsmodels.formula.api import ols
In [46]: model = ols("CA ~ NY", data=df).fit()
In [47]: print model.summary2()
              Results: Ordinary least squares
     ==========================================================
     Model:                OLS    AIC:               366.3982
     Dependent Variable:   CA     BIC:               369.2662
     No. Observations:     31     Log-Likelihood:    -181.20
     Df Model:             1      F-statistic:       94.26
     Df Residuals:         29     Prob (F-statistic): 1.29e-10
     R-squared:            0.765  Scale:             7473.2
     Adj. R-squared:       0.757
     ----------------------------------------------------------
                Coef.   Std.Err.   t    P>|t|    [0.025   0.975]
     ----------------------------------------------------------
     Intercept 65.7707  62.2894 1.0559 0.2997 -61.6254 193.1667
     NY         2.8213   0.2906 9.7085 0.0000   2.2270   3.4157
     ----------------------------------------------------------
     Omnibus:              1.502    Durbin-Watson:        0.921
     Prob(Omnibus):        0.472    Jarque-Bera (JB):     1.158
     Skew:                -0.465    Prob(JB):             0.560
     Kurtosis:             2.823    Condition No.:        860
     ==========================================================
```

An interesting method to express the fact that we would like to obtain a
formula of the variable CA with respect to the variable NY: CA ~ NY.
This comfortable syntax is possible thanks to the library `patsy` that
takes care of making all the pertinent interpretations and handling the
corresponding data behind the scenes.

The fit can be visualized with the graphic routine `plot_fit` from the submodule
`statsmodels.graphics.regressionplots`:

```
In [48]: from statsmodels.graphics.regressionplots import plot_fit
In [49]: plot_fit(model, 'NY'); \
   ....: plt.show()
```

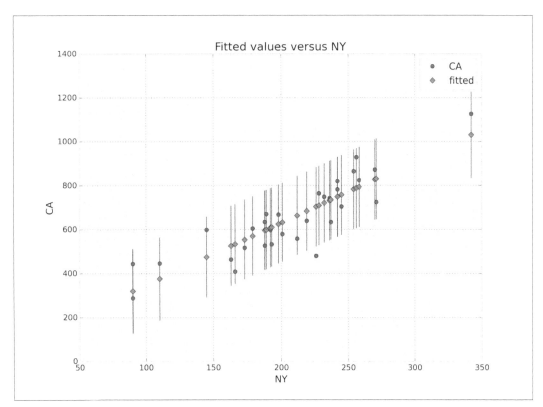

Let's also illustrate how to perform multiple linear regression with `statsmodels`. For the following example, we are going to gather the number of complaints during the year 2013 on three products that we suspect are related. We will try to find a formula that approximates the overall number of mortgage complaints as a function of the number of complaints on both credit cards and student loans:

```
In [50]: products = ['Student loan', 'Credit card', 'Mortgage']; \
   ....: products = data.Product.map(lambda t: t in products); \
   ....: df = data[products & in_2013]; \
   ....: df = df.groupby(['State', 'Product']).size()
   ....: df = df.unstack().dropna()
In [51]: X = df[['Credit card', 'Student loan']]; \
   ....: X = sm.add_constant(X); \
   ....: y = df['Mortgage']
In [52]: model = sm.OLS(y, X).fit(); \
   ....: print model.summary2()
                Results: Ordinary least squares
   =================================================================
   Model:                OLS          AIC:                 827.7591
   Dependent Variable:   Mortgage     BIC:                 833.7811
   No. Observations:     55           Log-Likelihood:      -410.88
   Df Model:             2            F-statistic:         286.6
   Df Residuals:         52           Prob (F-statistic):  8.30e-29
   R-squared:            0.917        Scale:               1.9086e+05
   Adj. R-squared:       0.914

   -----------------------------------------------------------------
              Coef.   Std.Err.    t      P>|t|    [0.025    0.975]
   -----------------------------------------------------------------
   const       2.1214  77.3360  0.0274 0.9782 -153.0648 157.3075
   Credit card 6.0196   0.5020 11.9903 0.0000    5.0121   7.0270
   Student loan -9.9299  2.5666 -3.8688 0.0003  -15.0802  -4.7796
   -----------------------------------------------------------------
   Omnibus:              20.251       Durbin-Watson:       1.808
   Prob(Omnibus):        0.000        Jarque-Bera (JB):    130.259
   Skew:                 -0.399       Prob(JB):            0.000
   Kurtosis:             10.497       Condition No.:       544
   =================================================================
```

Note the value of `r-squared`, so close to `1`. This indicates that a linear formula has been computed and the corresponding model fits the data well. We can now produce a visualization that shows it:

```
In [53]: from mpl_toolkits.mplot3d import Axes3D
In [54]: xspan = np.linspace(X['Credit card'].min(),
    ....:                     X['Credit card'].max()); \
    ....: yspan = np.linspace(X['Student loan'].min(),
    ....:                      X['Student loan'].max()); \
    ....: xspan, yspan = np.meshgrid(xspan, yspan); \
    ....: Z = model.params[0] + model.params[1] * xspan + \
    ....:     model.params[2] * yspan; \
    ....: resid = model.resid
In [55]: fig = plt.figure(figsize=(8, 8)); \
    ....: ax = Axes3D(fig, azim=-100, elev=15); \
    ....: surf = ax.plot_surface(xspan, yspan, Z, cmap=plt.cm.Greys,
    ....:                        alpha=0.6, linewidth=0); \
    ....: ax.scatter(X[resid>=0]['Credit card'],
    ....:            X[resid>=0]['Student loan'],
    ....:            y[resid >=0],
    ....:            color = 'black', alpha=1.0, facecolor='white'); \
    ....: ax.scatter(X[resid<0]['Credit card'],
    ....:            X[resid<0]['Student loan'],
    ....:            y[resid<0],
    ....:            color='black', alpha=1.0); \
    ....: ax.set_xlabel('Credit cards'); \
    ....: ax.set_ylabel('Student loans'); \
    ....: ax.set_zlabel('Mortgages'); \
    ....: plt.show()
```

The corresponding diagram shows data points above the plane in white and points below the plane in black. The intensity of the plane is determined by the corresponding predicted values for the number of mortgage complaints (brighter areas equal low residuals and darker areas equal to high residuals).

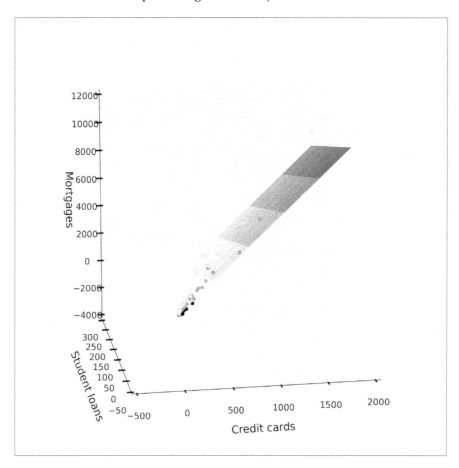

Ordinary least-squares regression for large datasets

In the case of linear regression for large datasets — more than 100,000 samples — optimal algorithms use **stochastic gradient descent (SGD)** regression learning with different loss functions and penalties. We can access these with the class SGDregressor in sklearn.linear_model.

Linear regression beyond ordinary least-squares

In the general case of multiple linear regression, if no emphasis is to be made on a particular set of variables, we employ ridge regression. We do so through the class `sklearn.linear_model.Ridge` in the `scikit-learn` toolkit.

Ridge regression is basically an ordinary least-squares algorithm with an extra penalty imposed on the size of the coefficients involved. It is comparable in performance to ordinary least-squares too, since they both have roughly the same complexity.

In any given multiple linear regression, if we acknowledge that only a few variables have a strong impact over the overall regression, the preferred methods are **Least Absolute Shrinkage and Selection Operator (LASSO)** and **Elastic Net**. We choose lasso when the number of samples is much larger than the number of variables, and we seek a sparse formula where most of the coefficients associated to non-important variables are zero.

Elastic net is always the algorithm of choice when the number of variables is close to, or larger than, the number of samples.

Both methods can be implemented through `scikit-learn`—the classes `sklearn.linear_model.lasso` and `sklearn.linear_model.ElasticNet`.

Support vector machines

Support vector regression is another powerful algorithm based upon the premise that a subset of the training data has a strong effect on the overall set of variables. An advantage of this method for unbalanced problems is that, by simply changing a kernel function in the algorithm as our decision function, we are able to access different kinds of regression (not only linear!). It is not a good choice when we have more variables than samples, though.

In the `scipy-learn` toolkit, we have two different classes that implement variations of this algorithm: `svm.SCV` and `svm.NuSVC`. A simplified variation of `svm.SVC` with linear kernel can be called with the class `svm.LinearSVC`.

Ensemble methods

When everything else fails, we have a few algorithms that combine the power of several base estimators. These algorithms are classified in two large families:

- **Averaging methods**: This combines estimators by averaging to reduce the value of the variance of the residuals

- **Boosting methods**: This builds a sequence of weak estimators that converge to a regression without any bias

For an in-detail description of these methods, examples, and implementation through the `scikit-learn` toolkit, refer to the official documentation at `http://scikit-learn.org/stable/modules/ensemble.html`.

Analysis of the time series

The subfield of time series modeling and analysis is also very rich. This kind of data arises in many processes, ranging from corporate business/industry metrics (economic forecasting, stock market analysis, utility studies, and many more), to biological processes (epidemiological dynamics, population studies, and many more). The idea behind modeling and analysis lies in the fact that data taken over time might have some underlying structure, and tracking that trend is desirable for prediction purposes, for instance.

We employ a **Box-Jenkins** model (also known as the **Autoregressive Integrated Moving Average** model, or **ARIMA** in short) to relate the present value of a series to past values and past prediction errors. In the package `statsmodels`, we have implementations through the following classes in the submodule `statsmodels.tsa`:

- To describe an **autoregressive** model **AR(p)**, we use the class `ar_model.AR`, or equivalently, `arima_model.AR`.

- To describe an **autoregressive Moving Average** model **ARMA(p,q)** we employ the class `arima_model.ARMA`. There is an implementation of the Kalman filter to aid with ARIMA models. We call this code with `kalmanf.kalmanfilter.KalmanFilter` (again, in the `statsmodels.tsa` submodule).

- For the general **ARIMA** model **ARIMA(p,d,q)**, we use the classes `arima_model.ARIMA`.

There are different classes of models that can be used to predict a time series from its own history—regression models that use lags and differences, random walk models, exponential smoothing models, and seasonal adjustment. In reality, all of these types of models are special cases of the Box-Jenkins models.

One basic example will suffice. Recall the time series we created by gathering all daily complaints in the years 2013 and 2014. From its plot, it is clear that our time series `ts` is not seasonal and not stationary (there is that slight upward trend). The series is too spiky for a comfortable analysis. We proceed to resample it weekly, prior to applying any description. We use, for this task, the resample method for the `pandas` time series:

```
In [56]: ts = ts.resample('W')
In [57]: ts.plot(); \
   ....: plt.show()
```

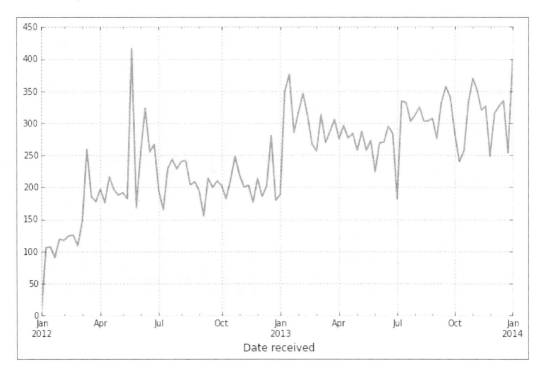

A first or second difference is used to detrend it—this indicates that we must use an ARIMA(p,d,q) model with `d=1` or `d=2` to describe it.

Let's compute and visualize the first differences series `ts[date] - ts[date-1]`:

 In pandas, we have a nice method for computing differences between any periods, for example, `ts.diff(periods=k)`.

```
In [58]: ts_1st_diff = ts.diff(periods=1)[1:]
In [59]: ts_1st_diff.plot(marker='o'); \
    ....: plt.show()
```

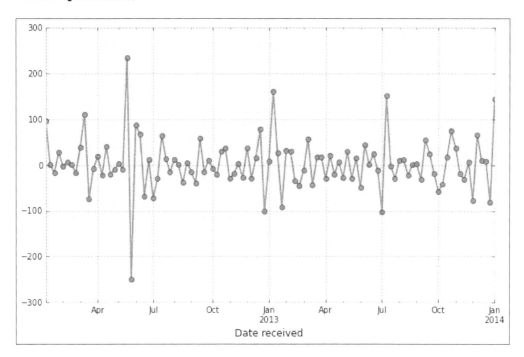

It indeed looks like a stationary series. We will thus choose `d=1` for our ARIMA model. Next, is the visualization of the correlograms of this new time series. We have two methods to perform this task: a basic autocorrelation and lag plot from `pandas` and a set of more sophisticated correlograms from `statsmodels.api.graphics.tsa`:

```
In [60]: from pandas.tools.plotting import autocorrelation_plot, \
    ....:                                 lag_plot
In [61]: autocorrelation_plot(ts_1st_diff); \
    ....: plt.show()
```

This gives us the following figure, showing the correlation of the data with itself at varying time lags (from 0 to 1000 days, in this case). The solid black line corresponds to the 95 percent confidence band, and the dashed line to the 99 percent confidence band:

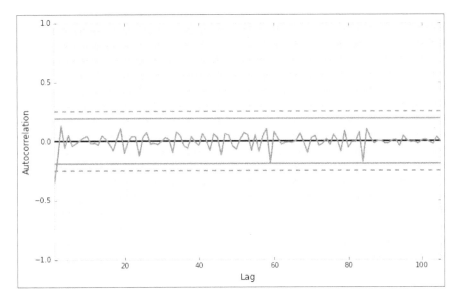

In the case of a nonrandom time series, one or more of the autocorrelations will be significantly nonzero. Note that, in our case, this plot is a good case for randomness.

The lag plot reinforces this view:

```
In [62]: lag_plot(ts_1st_diff, lag=1); \
    ....: plt.axis('equal'); \
    ....: plt.show()
```

In this case, we have chosen a lag plot of one day, which shows neither symmetry nor significant structure:

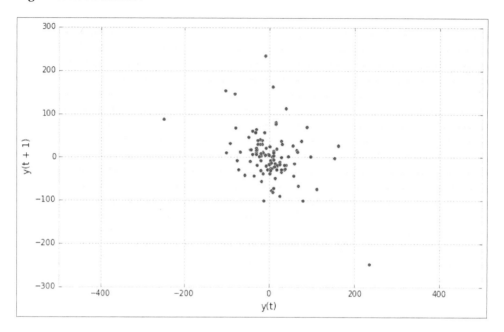

The statsmodels libraries excel in the treatment of time series through its submodules tsa and api.graphics.tsa. For example, to perform an autocorrelation as before, but this time restricting the lags to 40, we issue sm.graphics.tsa.plot_acf. We can use the following command:

```
In [63]: fig = plt.figure(); \
    ....: ax1 = fig.add_subplot(211); \
    ....: ax2 = fig.add_subplot(212); \
    ....: sm.graphics.tsa.plot_acf(ts_1st_diff, lags=40,
    ....:                          alpha=0.05, ax=ax1); \
    ....: sm.graphics.tsa.plot_pacf(ts_1st_diff, lags=40,
    ....:                          alpha=0.05, ax=ax2); \
    ....: plt.show()
```

This is a different way to present the autocorrelation function but in a way that is equally effective. Notice how we have control over the confidence band. Our choice of alpha determines its meaning—in this case, for example, by choosing `alpha=0.05`, we have imposed a 95 percent confidence band.

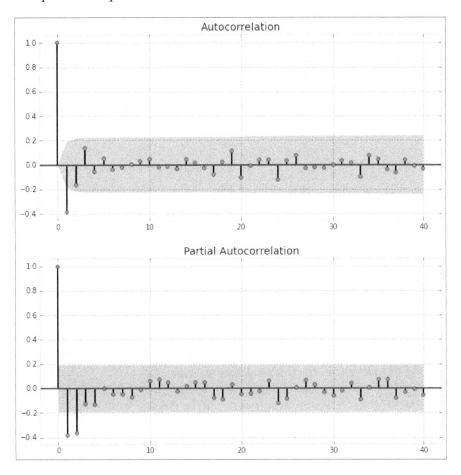

To present the corresponding values to the computed autocorrelations, we use the routines `acf` and `pacf` from the submodule `statsmodels.tsa.stattools`:

```
In [64]: from statsmodels.tsa.stattools import acf, pacf
In [65]: acf(ts_1st_diff, nlags=40)
Out[65]:
array([ 1.        , -0.38636166, -0.16209701,  0.13397057,
       -0.0555708 ,  0.05048394, -0.0407119 , -0.02082811,
```

```
       0.0040006 ,   0.02907198,   0.04330283,  -0.02011055,
      -0.01537464,  -0.02978855,   0.04849505,   0.01825439,
      -0.02593023,  -0.07966487,   0.02102888,   0.10951272,
      -0.10171504,  -0.00645926,   0.03973507,   0.03865624,
      -0.12395291,   0.03391616,   0.07447618,  -0.02474901,
      -0.01742892,  -0.02676263,  -0.00276295,   0.03135769,
       0.0155686 ,  -0.09556651,   0.07881427,   0.04804349,
      -0.03797063,  -0.05942366,   0.03913402,  -0.00854744, -
       0.03463874])
In [66]: pacf(ts_1st_diff, nlags=40)
Out[66]:
array([ 1.        ,  -0.39007667,  -0.37436812,  -0.13265923,
      -0.14290863,  -0.00457552,  -0.05168091,  -0.05241386,
      -0.07909324,  -0.01776889,   0.06631977,   0.07931566,
       0.0567529 ,  -0.02606054,   0.02271939,   0.05509316,
       0.06013166,  -0.09309867,  -0.11283787,   0.03704051,
      -0.06223677,  -0.05993707,  -0.03659954,   0.07764279,
      -0.16189567,  -0.11602938,   0.00638008,   0.09157757,
       0.04046057,  -0.04838127,  -0.08806197,  -0.02527639,
       0.06392126,  -0.13768596,   0.00641743,   0.11618549,
       0.12550781,  -0.14070774,  -0.05995693,  -0.0024937 ,
      -0.0905665 ])
```

In view of the correlograms computed previously, we have a MA(1) signature; there is a single negative spike in the ACF plot, and a decay pattern (from below) in the PACF plot. A sensible choice of parameters for the ARIMA model is then p=0, q=1, and d=1 (this corresponds to a simple exponential smoothing model, possibly with a constant term added). Let's then proceed with the model description and further forecasting, with this choice:

```
In [67]: from statsmodels.tsa import arima_model
In [68]: model = arima_model.ARIMA(ts, order=(0,1,1)).fit()
```

While running, this code informs us of several details of its implementation:

```
RUNNING THE L-BFGS-B CODE

           * * *

Machine precision = 2.220D-16
 N =            2    M =            12
 This problem is unconstrained.


At X0          0 variables are exactly at the bounds


At iterate    0    f=  5.56183D+02    |proj g|=  2.27373D+00


At iterate    5    f=  5.55942D+02    |proj g|=  1.74759D-01


At iterate   10    f=  5.55940D+02    |proj g|=  0.00000D+00


          * * *


Tit   = total number of iterations
Tnf   = total number of function evaluations
Tnint = total number of segments explored during Cauchy searches
Skip  = number of BFGS updates skipped
Nact  = number of active bounds at final generalized Cauchy point
Projg = norm of the final projected gradient
F     = final function value


          * * *


   N    Tit    Tnf  Tnint  Skip  Nact    Projg        F
   2    10     12     1     0     0    0.000D+00    5.559D+02
   F =   555.940373727761
```

```
CONVERGENCE: NORM_OF_PROJECTED_GRADIENT_<=_PGTOL


Cauchy                  time 0.000E+00 seconds.

Subspace minimization  time 0.000E+00 seconds.

Line search             time 0.000E+00 seconds.


Total User time 0.000E+00 seconds.
```

The method `fit` of the class `arima_model.ARIMA` creates a new class in `statsmodels.tsa`, `arima_model_ARIMAResults`, that holds all the information we need, and a few methods to extract it:

```
In [69]: print model.summary()
```

```
                              ARIMA Model Results
==============================================================================
Dep. Variable:                    D.y   No. Observations:                  105
Model:                 ARIMA(0, 1, 1)   Log Likelihood                -555.940
Method:                       css-mle   S.D. of innovations             48.047
Date:                Tue, 21 Oct 2014   AIC                           1117.881
Time:                        00:01:44   BIC                           1125.843
Sample:                    01-08-2012   HQIC                          1121.107
                         - 01-05-2014
==============================================================================
                 coef    std err          z      P>|z|      [95.0% Conf. Int.]
------------------------------------------------------------------------------
const          2.4303      1.330      1.827      0.071      -0.177      5.038
ma.L1.D.y     -0.7246      0.070    -10.341      0.000      -0.862     -0.587
                                     Roots
==============================================================================
                  Real          Imaginary           Modulus         Frequency
------------------------------------------------------------------------------
MA.1            1.3800           +0.0000j            1.3800            0.0000
------------------------------------------------------------------------------
```

Let's observe the correlograms of the residuals. We can compute those values using the method `resid` of the object model:

```
In [70]: residuals = model.resid
In [71]: fig = plt.figure(); \
   ....: ax1 = fig.add_subplot(211); \
   ....: ax2 = fig.add_subplot(212); \
   ....: sm.graphics.tsa.plot_acf(residuals, lags=40,
   ....:                          alpha=0.05, ax=ax1); \
```

```
    ....: ax1.set_title('Autocorrelation of the residuals of the
ARIMA(0,1,1) model'); \

    ....: sm.graphics.tsa.plot_pacf(residuals, lags=40,

    ....:                           alpha=0.05, ax=ax2); \

    ....: ax2.set_title('Partial Autocorrelation of the residuals of the
ARIMA(0,1,1) model'); \

    ....: plt.show()
```

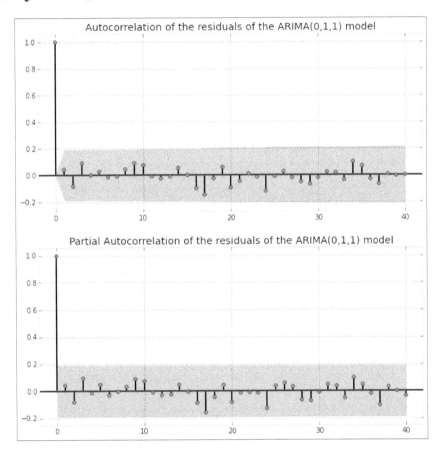

These plots suggest that we chose a good model. It only remains to produce a forecast using it. For this, we employ the method `predict` in the object model. For instance, a prediction for the first weeks in the year 2014, performed by considering all data since October 2013, could be performed as follows:

```
In [72]: np.where((ts.index.year==2013) & (ts.index.month==10))
Out[72]: (array([92, 93, 94, 95]),)
In [73]: prediction = model.predict(start=92, end='1/15/2014')
```

```
In [74]: prediction['10/2013':].plot(lw=2, label='forecast'); \
   ....: ts_1st_diff['9/2013':].plot(style='--', color='red',
   ....:                   label='True data (first differences)'); \
   ....: plt.legend(); \
   ....: plt.show()
```

This gives us the following forecast:

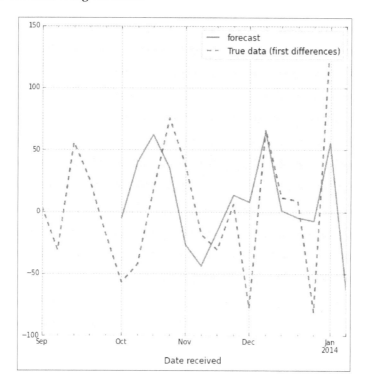

Summary

This concludes the first part of our two-chapter coverage of Data Analysis, where we have explored advanced Python tools in the SciPy stack for computation, and visualization of descriptive statistics. In the next chapter, we produce a similar treatment of inference statistics, data mining, and machine learning.

Inference and Data Analysis

8

The different techniques of descriptive statistics that we have covered in the previous chapter give us a straightforward presentation of facts from the data. The next logical step is **Inference** — the process of making propositions and drawing conclusions about a larger population than the sample data represents.

This chapter will cover the following topics:

- Statistical inference.
- Data mining and machine learning.

Statistical inference

Statistical inference is the process of deducing properties of an underlying distribution by analysis of data. Inferential statistical analysis infers properties about a population; this includes testing hypotheses and deriving estimates.

There are three types of inference:

- *Estimation* of the most appropriate single value of a parameter.
- *Interval estimation* to assess what region of parameter values is most consistent with the given data.
- *Hypothesis testing* to decide, between two options, what parameter values are most consistent with the data.

There are mainly three approaches to attack these problems:

- **Frequentist**: Inference is judged based upon performance in repeated sampling.

- **Bayesian**: Inference must be subjective. A prior distribution is chosen for the parameter we seek, and we combine the density of the data prior to obtain a joint distribution. A further application of Bayes Theorem gives us a distribution of the parameter, given the data. To perform computations in this setting, we use the package PyMC.

- **Likelihood**: Inference is based on the fact that all information about the parameter can be obtained by inspection of a likelihood function, which is proportional to the probability density function.

In this section, we briefly illustrate the three approaches to each of the three inference types. We go back to the previous example of ratios of daily complaints of mortgages against credit cards:

```
In [1]: import numpy as np, pandas as pd, matplotlib.pyplot as plt
In [2]: data = pd.read_csv("Consumer_Complaints.csv", \
   ...:                     low_memory=False, parse_dates=[8,9])
In [3]: df = data.groupby(['Date received', 'Product']).size(); \
   ...: df = df.unstack(); \
   ...: ratios = df['Mortgage'] / df['Credit card']
In [4]: ratios.describe()
Out[4]:
count    1001.000000
mean        2.939686
std         1.341827
min         0.203947
25%         1.985507
50%         2.806452
75%         3.729167
max        12.250000
dtype: float64
```

By visual inspection of the histogram, we could very well assume that this data is a random sample from a normal distribution with parameters mu (average) and sigma (standard deviation). We further assume, for simplicity, that the scale parameter sigma is known, and its value is 1.3.

Later in this section, we will actually explore what tools we have in the SciPy stack to determine the distribution of data more precisely.

Estimation of parameters

In this setting, the problem we want to solve is an estimation of the average, mu, using the data obtained.

Frequentist approach

This is the simplest setting. The frequentist approach uses as estimate the computed mean of the data:

```
In [5]: ratios.mean()
Out[5]: 2.9396857495543731
In [6]: from scipy.stats import sem   # Standard error
In [7]: sem(ratios.dropna())
Out[7]: 0.042411109594665049
```

 A frequentist would then say: *The estimated value of the parameter* mu *is* 2.9396857495543731 *with standard error* 0.042411109594665049.

Bayesian approach

For the Bayesian approach, we select a prior distribution for mu, which we conveniently assume is Normal with standard deviation 1.3. The average mu is regarded as a variable, and initially, we assume that its value could be anywhere in the range of the data (with Uniform distribution). We then use Bayes theorem to compute a **posterior distribution** for mu. Our estimated parameter is then the average of the posterior distribution of mu:

```
In [8]: import pymc as pm
In [9]: mu = pm.Uniform('mu', lower=ratios.min(), upper=ratios.max())
In [10]: observation = pm.Normal('obs', mu=mu, tau=1./1.3**2,
    ....:                        value=ratios.dropna(), observed=True)
In [11]: model = pm.Model([observation, mu])
```

Notice how, in PyMC, the definition of a Normal distribution requires an average parameter mu, but instead of standard deviation or variance, it expects the **precision** tau = 1/sigma**2.

The variable observation combines our data with our proposed data-generation scheme, given by the variable mu, through the option value=ratios.dropna(). To make sure that this stays fixed during the analysis, we impose observed=True.

In the learning step, we employ the **Markov Chain Monte Carlo (MCMC)** method to return a large amount of random variables for the posterior distribution of mu:

```
In [12]: mcmc = pm.MCMC(model)
In [13]: mcmc.sample(40000, 10000, 1)
[---------------100%---------------] 40000 of 40000 complete in 4.5 sec
In [14]: mcmc.stats()
Out[14]:
{'mu': {'95% HPD interval': array([ 2.86064764,  3.02292213]),
  'mc error': 0.00028222883254203107,
  'mean': 2.9396811517572554,
  'n': 30000,
  'quantiles': {2.5: 2.8589908555161485,
   25: 2.9117191652137464,
   50: 2.9396815504225815,
   75: 2.9675088640073439,
   97.5: 3.0216312862055279},
  'standard deviation': 0.041412844137324857}}
In [15]: mcmc.summary()
Out[15]:
mu:
```

Mean	SD	MC Error	95% HPD interval

| 2.94 | 0.041 | 0.0 | [2.861 3.023] |

Posterior quantiles:

| 2.5 | 25 | 50 | 75 | 97.5 |

```
|---------------|===============|===============|---------------|
```

| 2.859 | 2.912 | 2.94 | 2.968 | 3.022 |

```
In [16]: from pymc.Matplot import plot as mcplot
In [17]: mcplot(mcmc); \
    ....: plt.show()
Plotting mu
```

We should get an output similar to the following:

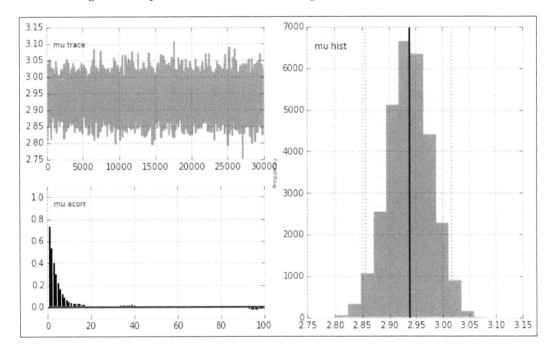

The estimated value of the parameter is 2.93968. The standard deviation of the posterior distribution of mu is 0.0414.

Likelihood approach

We have a convenient method to perform the likelihood approach for estimation of parameters of any distribution represented as a class in the submodule `scipy.stats`. In our case, since we are fixing the standard deviation (the `scale` as parameter of the normal distribution for this particular class), we would issue the following command:

```
In [18]: from scipy.stats import norm as NormalDistribution
In [19]: NormalDistribution.fit(ratios.dropna(), fscale=1.3)
Out[19]: (2.9396857495543736, 1.3)
```

This gives us a similar value for the mean. The graph of the (non-negative log) likelihood function for mu can be obtained as follows:

```
In [20]: nnlf = lambda t: NormalDistribution.nnlf([t, 1.3],
    ....:                                 ratios.dropna()); \
    ....: nnlf = np.vectorize(nnlf)
In [21]: x = np.linspace(0, 14); \
    ....: plt.plot(x, nnlf(x), lw=2, color='r',
    ....:         label='Non-negative log-likely function for $\mu$'); \
    ....: plt.legend(); \
    ....: plt.annotate('Minimum', xy=(2.9, nnlf(2.9)), xytext=(0,20),
    ....:         textcoords='offset points', ha='right', va='bottom',
    ....:         bbox=dict(boxstyle='round,pad=0.5', fc='yellow',
    ....:                 color='k', alpha=1),
    ....:         arrowprops=dict(arrowstyle='->', color='k',
    ....:                 connectionstyle='arc3,rad=0')); \
    ....: plt.show()
```

We should get an output similar to the following:

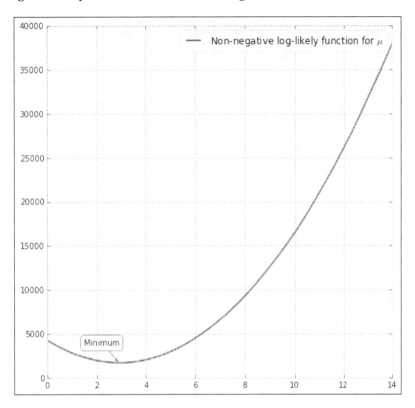

In any case, the result is visually what we would expect:

```
In [22]: distribution = NormalDistribution(loc=2.9396857495543736,
    ....:                                   scale=1.3)
In [23]: plt.plot(x, distribution.pdf(x), 'r-', lw=2,
    ....:              label='Computed Probability Density Function'); \
    ....: ratios.hist(bins=50, alpha=0.2, normed=True,
    ....:              label='Histogram of data (normalized)'); \
    ....: plt.legend(); \
    ....: plt.show()
```

We should get an output similar to the following:

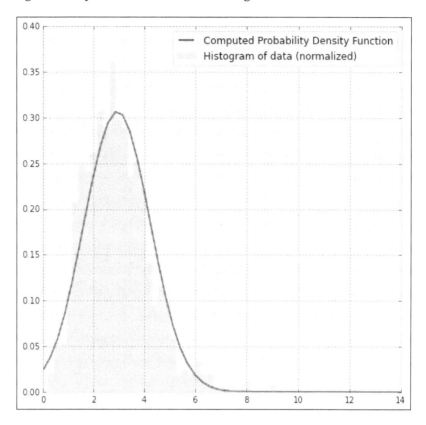

Interval estimation

In this setting, we seek an interval of values for mu that are supported by the data.

Frequentist approach

In the frequentist approach, we start by providing a small **confidence coefficient** alpha, and proceed to find an interval so that the probability of including the parameter mu is 1-alpha. In our example, we set alpha = 0.05 (hence, the probability we impose is 95%), and proceed to compute the interval with the method interval of any class defining a continuous distribution in the module scipy.stats:

```
In [24]: loc = ratios.mean(); \
    ....: scale = ratios.sem(); \
    ....: NormalDistribution.interval(0.95, scale=scale, loc=loc)
Out[24]: (2.8565615022044484, 3.0228099969042979)
```

According to this method, values of the average `mu` between `2.8565615022044484` and `3.0228099969042979` are consistent with the data based on a 95% confidence interval.

Bayesian approach

In the Bayesian approach, the equivalent to the confidence interval is called a **credible region** (or **interval**), and is associated with the **highest posterior density region**—the set of most probable values of the parameter that, in total, constitute `100*(1 - alpha)` percentage of the posterior mass.

Recall that when using a MCMC, after sampling, we obtained the credible region for `alpha = 0.05`.

To obtain credible intervals for other values of alpha, we use the routine `hpd` in the submodule `pymc.utils` directly. For example, the highest posterior density region for `alpha = 0.01` is computed as follows:

```
In [25]: pm.utils.hpd(mcmc.trace('mu')[:], 1-.99)
Out[25]: array([ 2.83464531,  3.04706652])
```

Likelihood approach

This is also done with the aid of the method `nnlf` of any distribution. In this setting, we need to determine the interval of parameter values for which the likelihood exceeds `1/k` where `k` is either 8 (strong evidence) or 32 (very strong evidence).

The estimation of the corresponding interval is then a simple application of optimization. We leave this as an exercise.

Data mining and machine learning

We are going to focus on three kinds of problems: *Classification*, *Dimensionality reduction*, and *Clustering*. Each of these problems is used in both data mining and machine learning to draw conclusions about the data. Let's explain each of these settings in different sections.

Classification

Classification is an example of **supervised learning**. There is a set of **training data** with an attribute that classifies it in one of several categories. The goal is to find the value of that attribute for new data. For example, with our running database, we could use all the data from the year 2013 to figure out which financial complaints got solved positively for the customer, which ones got solved without relief, and which ones remained *in progress*. This will offer us good insight on, for instance, which companies are faster to respond to consumer complaints positively, if there are states where complaints are less likely to get resolved, and so on.

Let's start by finding the kind of company responses observed in the database:

```
In [1]: import numpy as np, pandas as pd, matplotlib.pyplot as plt
In [2]: data = pd.read_csv("Consumer_Complaints.csv",
   ...:                         low_memory=False, parse_dates=[8,9])
In [3]: print data['Company response'].unique()
['Closed with non-monetary relief' 'In progress' 'Closed with
explanation'
 'Closed with monetary relief' 'Closed' 'Untimely response'
 'Closed without relief' 'Closed with relief']
```

These are eight different categories and our target for deciding the fate of future complaints. Let's create a set of training data by gathering all complaints formulated during the year 2013, and keeping only the columns that we believe are relevant to the decision-making process:

- The product and subproduct that originate the complaint.
- The issue (but not the sub-issue) that consumers had with the product.
- State (but not ZIP code) where the complaint was filed.
- Method of submission of the complaint.
- The company that offered the service.

The size of this training data will dictate which algorithm is to be used for classification.

Before the data can be processed, we need to encode nonnumerical labels so they can be properly treated by the different classifying algorithms. We do that with the class LabelEncoder from the module sklearn.preprocessing.

We will then try to classify all complaints formulated during the year 2014:

```
In [4]: in_2013 = data['Date received'].map(lambda t: t.year==2013);\
   ...: in_2014 = data['Date received'].map(lambda t: t.year==2014);\
   ...: df = data[in_2013 | in_2014]; \
   ...: df['Year'] = df['Date received'].map(lambda t: t.year); \
   ...: irrelevant = ['Date received', 'Date sent to company',
   ...:               'Complaint ID', 'Timely response?',
   ...:               'Consumer disputed?', 'Sub-issue','ZIP code'];\
   ...: df.drop(irrelevant, 1, inplace=True); \
   ...: df = df.dropna()
In [5]: from sklearn.preprocessing import LabelEncoder
In [6]: encoder = {}
In [7]: for column in df.columns:
   ...:     if df[column].dtype != 'int':
   ...:         le = LabelEncoder()
   ...:         le.fit(df[column].unique())
   ...:         df[column] = le.transform(df[column])
   ...:         encoder[column] = le
   ...:
In [8]: training = df[df.Year==2013]; \
   ...: target = training['Company response']; \
   ...: training.drop(['Company response', 'Year'], 1, inplace=True)
In [9]: test = df[df.Year==2014]; \
   ...: true_result = test['Company response']; \
   ...: test.drop(['Company response', 'Year'], 1, inplace=True)
In [10]: len(training)
Out[10]: 77100
```

Support vector classification

The training data here is not too big (anything less than 100,000 is considered manageable). For this volume of training data, it is suggested that we employ support vector classification with a linear kernel.

Three flavors of this algorithm are coded as classes in the module `sklearn.svm` (for support vector machines): SVC, NuSVC, and a simplified version of SVC with linear kernel, LinearSVC, which is what we need:

```
In [11]: from sklearn.svm import LinearSVC
In [12]: clf = LinearSVC(); \
   ....: clf.fit(training, target)
Out[12]:
LinearSVC(C=1.0, class_weight=None, dual=True, fit_intercept=True,
    intercept_scaling=1, loss='l2', multi_class='ovr', penalty='l2',
    random_state=None, tol=0.0001, verbose=0)
```

We are ready to evaluate the performance of this classifier:

```
In [13]: clf.predict(test)==true_result
Out[13]:
0        False
2        False
3         True
4         True
6        False
7         True
9         True
11       False
12       False
13       False
14       False
15        True
16        True
20       False
21       False
...
101604      True
```

```
101607      False
101610      True
101611      True
101613      True
101614      True
101616      True
101617      True
101618      True
101620      True
101621      True
101622      True
101625      True
101626      True
101627      True
Name: Company response, Length: 65282, dtype: bool
In [18]: float(sum(_)) / float(len(_))
Out[18]: 0.7985509022395147
```

With this method, we correctly classify almost 80% of the complaints.

 In rare cases where this method does not work, we can always resort to plain SVC or its NuSVC variation with a carefully chosen kernel.

The power of a classifier lies in the applications. For instance, if we would like to purchase via web a *conventional fixed mortgage* from Bank of America in the state of South Carolina, and we fear problems with *settlement process and cost*, what does the classifier tell us about our chances of having the matter settled?

```
In [19]: encoder['Product'].transform(['Mortgage'])[0]
Out[19]: 4
In [20]: encoder['Sub-product'].transform(['Conventional fixed
mortgage'])[0]
Out[20]: 5
In [21]: encoder['Issue'].transform(['Settlement process and costs'])[0]
Out[21]: 27
In [19]: encoder['State'].transform(['SC'])[0]
Out[19]: 50
In [23]: encoder['Submitted via'].transform(['Web'])[0]
Out[23]: 5
```

```
In [24]: encoder['Company'].transform(['Bank of America'])[0]
Out[24]: 247
In [25]: clf.predict([4,5,27,50,5,247])
Out[25]: array([1])
In [26]: encoder['Company response'].inverse_transform(_)[0]
Out[26]: 'Closed with explanation'
```

A satisfactory outlook!

Trees

It is possible to create a decision tree illustrating a set of rules to facilitate the classification. In the `scikit-learn` toolkit, we have a class implemented for this purpose—`DecisionTreeClassifier` in the submodule `sklearn.tree`. Let's see it in action:

```
In [27]: from sklearn.tree import DecisionTreeClassifier
In [28]: clf = DecisionTreeClassifier().fit(training, target)
In [29]: clf.predict(test) == true_result
Out[29]:
0        True
2        False
3        True
4        True
6        False
7        True
9        True
11       False
12       False
13       False
14       False
15       True
16       True
20       False
21       False
...
101604      True
```

```
101607      True
101610      True
101611      True
101613      True
101614      True
101616      True
101617      True
101618      True
101620      True
101621      True
101622      True
101625      True
101626     False
101627      True
Name: Company response, Length: 65282, dtype: bool
In [30]: float(sum(_)) / len(_)
Out[30]: 0.7400661744431849
```

It looks like this simple classifier was successful in predicting the outcome of about 74% of the complaints in 2014.

 It is possible to create a dot file readable with the Graphviz visualization software available at http://www.graphviz.org/:

```
In [31]: from sklearn.tree import export_graphviz
In [32]: export_graphviz(clf, out_file="tree.dot")
```

Opening this file in Graphviz gives us an impressive set of rules

The following is a detail of the tree (too large to fit in these pages!):

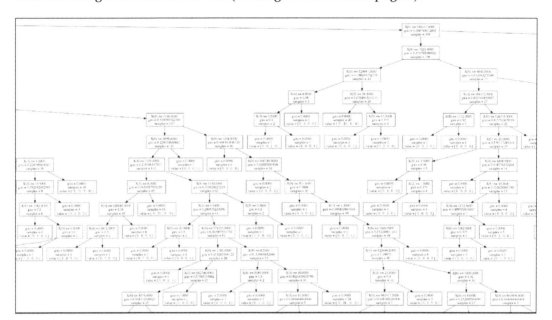

We also have implementations of random forests and extremely randomized trees, both of them within the submodule `sklearn.ensemble`. The corresponding classes are called `RandomForestClassifier` and `ExtraTreesClassifier`, respectively.

In any of the previous cases, the coding of the classifier is exactly as in the cases of the SVC and the basic decision trees.

Naive Bayes

Similar results are obtained with Naive Bayes methods. In the module `sklearn.naive_bayes`, we have three implementations of this algorithm:

- The class `GaussianNB` for the Gaussian Naive Bayes, where the likelihood of the features is assumed to be Gaussian.

- The class `BernoulliNB` for Naive Bayes for data distributes according to multivariate Bernoulli distributions (each feature is assumed to be a binary valued variable).

- The class `MultinomialNB` for multinomially distributed data.

Nearest neighbors

For an even better result for this case, we employ classification by nearest neighbors. This is exactly the same procedure we employed in the setting of computational geometry to perform the corresponding geometric query problem. In this setting, note how we have coded our data as points in a Euclidean space of high dimension, and we can thus translate those methods for this classification purpose.

The advantage in this case is that we don't necessarily have to use Euclidean distances for our computations. For instance, since the data is essentially different regardless of its numerical value, it makes sense to impose a Hamming metric to calculate distance between labels. We have a generalization of the nearest neighbors algorithm implemented as the class KNeighborsClassifier in the module sklearn.neighbors:

```
In [33]: from sklearn.neighbors import KNeighborsClassifier
In [34]: clf = KNeighborsClassifier(n_neighbors=8,metric='hamming');\
    ....: clf.fit(training, target)
Out[34]:
KNeighborsClassifier(algorithm='auto', leaf_size=30,metric='hamming',
          n_neighbors=8, p=2, weights='uniform')
In [35]: clf.predict(test)==true_result
Out[35]:
0       True
2       False
3       True
4       True
6       False
7       True
9       True
11      False
12      False
13      False
14      False
15      True
16      True
20      False
```

```
21      False
...
101604       True
101607      False
101610       True
101611       True
101613       True
101614       True
101616       True
101617       True
101618       True
101620       True
101621       True
101622       True
101625       True
101626       True
101627       True
Name: Company response, Length: 65282, dtype: bool
In [36]: float(sum(_))/len(_)
Out[36]: 0.791274777120799
```

More than 79% of success!

Dimensionality reduction

Data often observes internal structure, but high dimension (number of columns, in a sense) makes it difficult to extract and select this internal structure. Often, it is possible to perform smart projections of this data on lower-dimensional manifolds, and analyze these projections for search of features. We refer to this technique as **dimensionality reduction**.

For the following example, we decided to drop all dates that contain any NaN. This significantly reduced the volume of the data, making the subsequent study and results simpler to understand. For a more elaborate and complete study, force all occurrences of NaN to be zeros — substitute the method dropna() with fillna(0).

Let's observe how to profit from these processes with our running example. We gather all the daily complains by product, and analyze the data:

```
In [37]: df = data.groupby(['Date received', 'Product']).size(); \
   ....: df = df.unstack().dropna()
In [38]: df.head()
Out[38]:
```

Product	Bank account or service	Consumer loan	Credit card \
Date received			
2013-11-06	66	14	41
2013-11-07	44	11	33
2013-11-08	49	11	36
2013-11-09	9	4	20
2013-11-11	15	4	23

Product	Credit reporting	Debt collection	Money transfers	Mortgage \
Date received				
2013-11-06	62	129	2	153
2013-11-07	43	99	2	128
2013-11-08	44	83	8	113
2013-11-09	19	33	2	23
2013-11-11	32	68	2	46

Product	Payday loan	Student loan
Date received		
2013-11-06	2	14
2013-11-07	8	10
2013-11-08	12	7
2013-11-09	3	4
2013-11-11	2	14

```
In [39]: df.shape
Out[39]: (233, 9)
```

We may regard this data as 233 points in a space of 9 dimensions.

Principal component analysis

For this small kind of data, and without any other prior information, one of the best procedures of dimensionality reduction results in projecting over a two-dimensional plane. However, it's not just any plane—we seek one projection that ensures that the projected data has the largest possible variance. We accomplish this with the information we obtain from eigenvalues and eigenvectors of a matrix that represents our data. This process is called **principal component analysis (PCA)**.

PCA is regarded as one of the most useful techniques in Statistical methods. For an amazing survey of theory (in both scopes of Linear Algebra and Statistics), coding techniques, and applications, the best resource is the second edition of the book *Principal Component Analysis*, written by I.T. Jolliffe and published by *Springer* in their *Springer Series in Statistics* in 2002.

We have an implementation in the `scikit-learn` toolkit, the class `PCA`, in the submodule `sklearn.decomposition`:

```
In [40]: from sklearn.decomposition import PCA
In [41]: model = PCA(n_components=2)
In [42]: model.fit(df)
Out[42]: PCA(copy=True, n_components=2, whiten=False)
In [43]: projected_df = model.transform(df)
In [44]: plt.figure(); \
   ....: plt.scatter(projected_df[:,0], projected_df[:,1]); \
   ....: plt.title('Principal Component Analysis scatterplot of df'); \
   ....: plt.show()
```

Observe how the data consists of two very well differentiated clusters of points—one of them considerably larger than the other—and some outliers. We will come back to this problem in the next section.

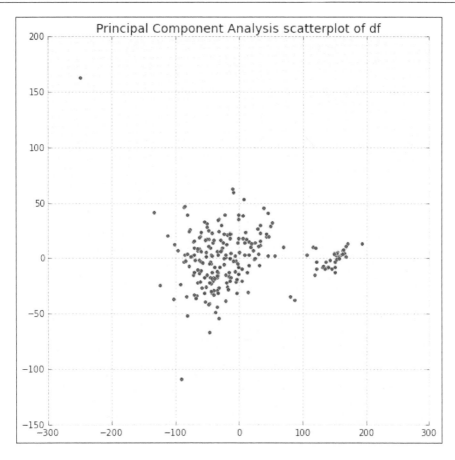

Isometric mappings

We don't necessarily need to project on hyperplanes. One neat trick is to assume that the data itself lies on a nonlinear submanifold, and obtain a representation of this object with the points on it. This gives us flexibility to search for projections where the projected data satisfies relevant properties. For instance, if we require the projections to maintain geodesic distance among points (whenever possible), we achieve a so-called **isometric mapping** (**isomap**).

In the SciPy stack, we have an implementation of this method as the class `Isomap` in the submodule `sklearn.manifold`:

```
In [45]: from sklearn.manifold import Isomap
In [46]: model = Isomap().fit(df)
In [47]: isomapped_df = model.transform(df)
In [48]: plt.figure(); \
```

```
....: plt.scatter(isomapped_df[:,0], isomapped_df[:,1]); \
....: plt.title('Isometric Map scatterplot of df'); \
....: plt.show()
```

Although visually very different, this method also offers us two very clear clusters, one of them much larger than the other. The smaller cluster appears as a sequence of points clearly aligned:

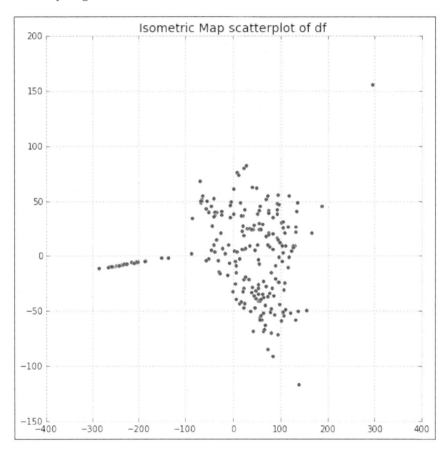

Spectral embedding

Another possibility is to embed the data nonlinearly by applying spectral analysis on an affinity/similarity matrix. The results carry similar quality as the previous two examples:

```
In [49]: from sklearn.manifold import SpectralEmbedding
In [50]: model = SpectralEmbedding().fit(df)
In [51]: embedded_df = model.embedding_
```

```
In [52]: plt.figure(); \
    ....: plt.scatter(embedded_df[:,0], embedded_df[:,1]); \
    ....: plt.title('Spectral Embedding scatterplot of df'); \
    ....: plt.show()
```

In this case, the clusters are more clearly defined than in the previous examples.

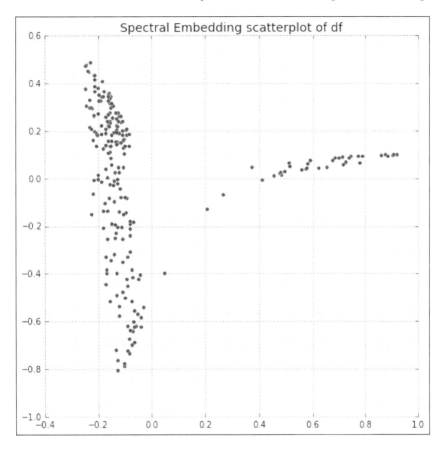

Locally linear embedding

Another possible projection, similar in some sense to isometric maps, seeks to preserve the distance within local neighborhoods — the **locally linear embedding**. We have an implementation through the class `LocallyLinearEmbedding`, again within the submodule `sklearn.manifold`:

```
In [53]: from sklearn.manifold import LocallyLinearEmbedding

In [54]: model = LocallyLinearEmbedding().fit(df)

In [55]: lle_df = model.transform(df)
```

```
In [56]: plt.figure(); \
   ....: plt.scatter(lle_df[:,0], lle_df[:,1]); \
   ....: plt.title('Locally Linear Embedding scatterplot of df'); \
   ....: plt.show()
```

Note the extremely conglomerated two clusters, and two outliers.

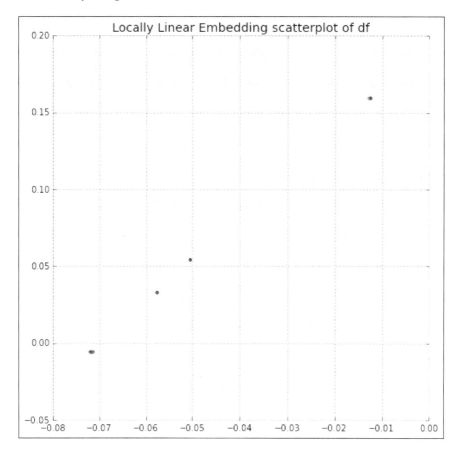

Clustering

Clustering is similar in some sense to the problem of classification, yet more complex. When facing a dataset, we acknowledge the possibility of having some hidden structure, in a way that will allow us to predict the behavior of future data. Searching for this structure is performed by finding common patterns and gathering data conforming to those patterns in different clusters. For this reason, we also refer to this problem as **data mining**.

There are many different methods to perform clustering, depending on the volume of data, and the a priori information we have on the number of clusters. We are going to explore the following settings:

- MeanShift.

- Gaussian mixture models.

- K-means.

- Spectral clustering.

MeanShift

We employ the technique of **MeanShift** clustering when the data does not exceed 10,000 points, and we do not know a priori the number of clusters we need. Let's experiment with the running example from the section on dimensionality reduction, but we will remain oblivious from the two clusters suggested by all projections. We will let the mean shift clustering take that decision for us.

We have an implementation in the SciPy stack through the class `MeanShift` in the submodule `sklearn.cluster` of the `scikit-learn` toolkit. One of the ingredients of this algorithm is approximation with radial basis functions (discussed in *Chapter 1, Numerical Linear Algebra*), for which we need to provide an appropriate bandwidth. The algorithm, if not provided with one, will try to estimate from the data. This process could potentially be very slow and expensive, and it is generally a good idea to do estimation by ourselves, so we can control resources. We can do so with the helper function `estimate_bandwidth`, in the same submodule.

 The implementation of classes and routines for clustering algorithms in the `scikit-learn` toolkit require the data to be fed as a `numpy` array, rather than a `pandas` data frame.

We can perform this switch easily with the `dataframe` method `.values`.

```
In [57]: from sklearn.cluster import MeanShift, estimate_bandwidth

In [58]: bandwidth = estimate_bandwidth(df.values, n_samples=1000)

In [59]: model = MeanShift(bandwidth=bandwidth, bin_seeding=True)

In [60]: model.fit(df.values)
```

At this point, the object `model` has successfully computed a series of labels and attached them to each point in the data, so they are properly clustered. We can allow some *unclassifiable* points to remain unclassified — we accomplish this by setting the optional Boolean flag `cluster_all` to `False`. By default, the algorithm forces every single piece of data into a cluster.

Let's find the number of labels and visualize the result on top of one of the projections from the previous section, for quality purposes:

```
In [61]: np.unique(model.labels_)          # how many clusters?
Out[61]: array([0,   1,   2])
In [62]: plt.figure(); \
    ....: plt.scatter(isomapped_df[:,0], isomapped_df[:,1],
    ....:             c=model.labels_, s = 50 + 100*model.labels_); \
    ....: plt.title('MeanShift clustering of df\n Isometric Mapping \
    ....: scatterplot\n color/size indicates cluster'); \
    ....: plt.tight_layout(); \
    ....: plt.show()
```

Note how the two clear clusters are correctly computed and one outlier received its own cluster.

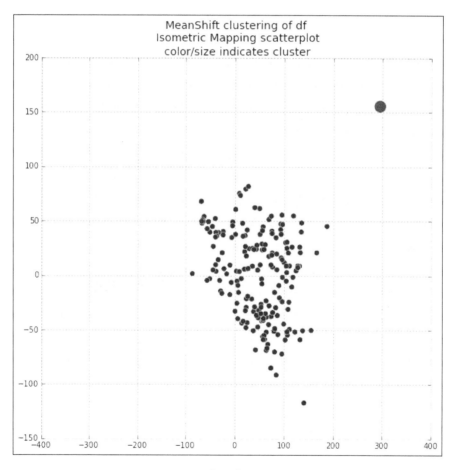

Let's find out the significance of these clusters. First, the outlier:

```
In [63]: df[model.labels_ == 2]
Out[63]:
Product        Bank account or service  Consumer loan  Credit card  \
Date received
2014-06-26                         117             19           89

Product        Credit reporting  Debt collection  Money transfers
Mortgage  \
Date received
2014-06-26                   85              159                5
420

Product        Payday loan  Student loan
Date received
2014-06-26               7            12
```

What is so different in the amount of complaints produced on July 26th, 2014? Let us produce a plot with the each cluster of dates, to see if we may guess what the differences are about.

```
In [64]: fig = plt.figure(); \
   ....: ax1 = fig.add_subplot(211); \
   ....: ax2 = fig.add_subplot(212); \
   ....: df[model.labels_==0].plot(ax=ax1); \
   ....: df[model.labels_==1].plot(ax=ax2); \
   ....: plt.show()
```

We should get an output similar to the following:

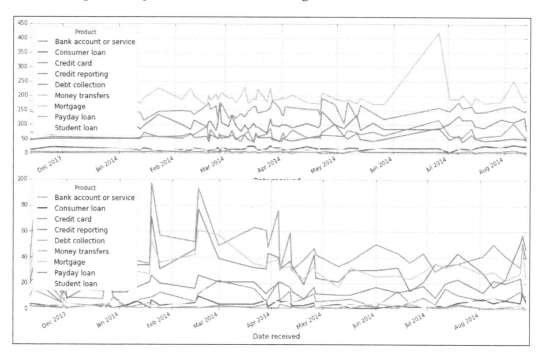

Visually, it appears as if the cluster has been formed by gathering dates with high volume of complaints, versus low volume of complaints. But not only that: a closer inspection reveals that on dates from cluster 0, mortgages are unequivocally the number one reason for complaint. On the other hand, for dates in cluster 1, complaints on mortgages get relegated to a second or third position, always behind debt collection and payday loans:

```
In [65]: plt.figure(); \
    ....: df[model.labels_==0].sum(axis=1).plot(label='cluster 0'); \
    ....: df[model.labels_==1].sum(axis=1).plot(label='cluster 1'); \
    ....: plt.legend(); \
    ....: plt.show()
```

Indeed, that was the case:

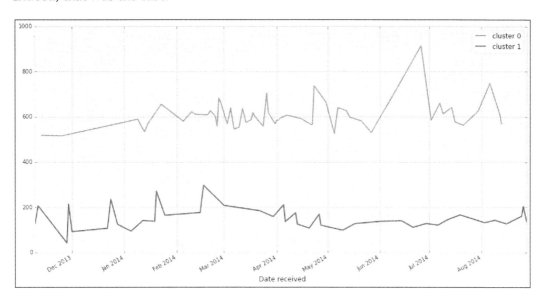

```
In [66]: df[model.labels_==0].sum(axis=1).describe()
Out[66]:
count    190.000000
mean     528.605263
std       80.900075
min      337.000000
25%      465.000000
50%      537.000000
75%      585.000000
max      748.000000
dtype: float64
In [67]: df[model.labels_==1].sum(axis=1).describe()
Out[67]:
count     42.000000
mean     156.738095
std       56.182655
min       42.000000
25%      124.250000
50%      140.500000
75%      175.750000
```

```
max          335.000000
dtype: float64
In [68]: df[model.labels_==2].sum(axis=1)
Out[68]:
Date received
2014-06-26          913
dtype: float64
```

Note that the volume of complaints in the cluster labeled 1 does not go over 335 daily complaints. Complaints formulated on days from the zero cluster are all between 337 and 748. On the outlier date — July 26th, 2014 — there were 913 complaints.

Gaussian mixture models

Gaussian mixture models are probabilistic models that make assumptions on the way the data has been generated, and the distributions it obeys. These algorithms approximate the parameters defining the involved distributions.

In its purest form, this method implements the **expectation-maximization (EM)** algorithm in order to fit the model. We access this implementation with the class GMM in the submodule `sklearn.mixture` of the `scikit-learn` toolkit. This implementation requires us to provide with the number of desired clusters, though. And unlike other methods, it will try its hardest to categorize the data into as many clusters required, no matter whether these artificial clusters make any logical sense.

In order to perform clustering on relatively small amounts of data without previous knowledge of the number of clusters that we need, we may employ a variant of Gaussian mixture models that use variational inference algorithms instead. We call this a **variational Gaussian mixture**. We have an implementation of this algorithm as the class VBGMM in the same submodule.

For this particular method, we do need to feed an upper bound of the number of clusters we expect, but the algorithm will compute the optimal number for us.

For instance, in our running example — which clearly shows two clusters — we are going to impose an upper bound of 30, and observe the behavior of the VBGMM algorithm:

```
In [69]: from sklearn.mixture import VBGMM
In [70]: model = VBGMM(n_components=30).fit(df)
In [71]: labels = model.predict(df)
In [72]: len(np.unique(labels))    # how many clusters?
Out[72]: 2
```

Only two clusters!

```
In [73]: a, b = np.unique(labels)
In [74]: sizes = 50 + 100 * (labels - a) / float(b-a)
In [75]: plt.figure(); \
    ....: plt.scatter(embedded_df[:,0], embedded_df[:,1],
    ....:                c=labels, s=sizes); \
    ....: plt.title('VBGMM clustering of df\n Spectral Embedding \
    ....: scatterplot\n color/size indicates cluster'); \
    ....: plt.tight_layout(); \
    ....: plt.show()
```

We should get an output similar to the following:

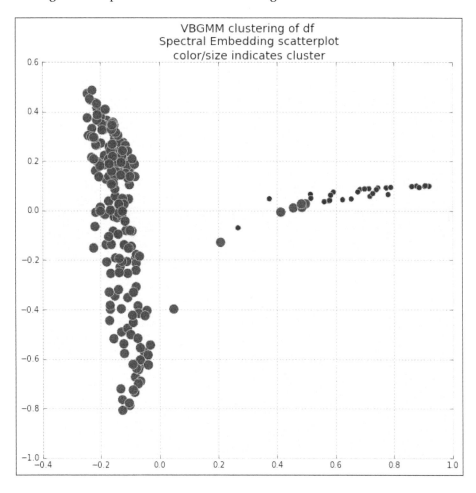

Kmeans

If we previously know the number of clusters that we require, regardless of the amount of data, a good algorithm for clustering is Lloyd's algorithm (better known as the method of K-means).

In the module `scipy.cluster.vq` we have an efficient set of routines for k-means clustering. A parallel-capable algorithm is implemented as the class `KMeans` in the submodule `sklearn.cluster` of the `scikit-learn` toolkit. For instance, if we require on our data a partition into four clusters, using all CPUs of our computer, we could issue from the toolkit the following code:

```
In [76]: from sklearn.cluster import KMeans
In [77]: model = KMeans(n_clusters=4, n_jobs=-1).fit(df)
In [78]: plt.figure(); \
    ....: plt.scatter(isomapped_df[:,0], isomapped_df[:,1],
    ....:                 c=model.labels_, s=10 + 100*model.labels_); \
    ....: plt.title('KMeans clustering of df\n Isometric Mapping \
    ....: scatterplot\n color/size indicates cluster'); \
    ....: plt.tight_layout(); \
    ....: plt.show()
```

We should get an output similar to the following:

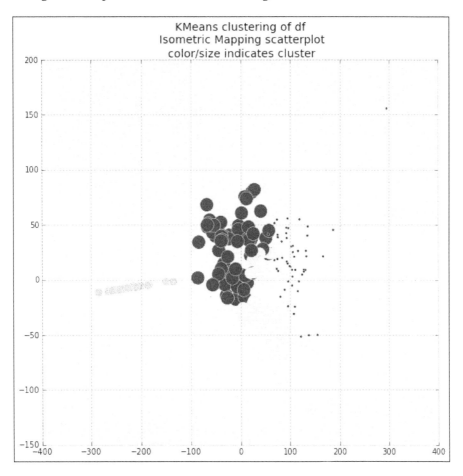

Note how this artificial clustering still manages to categorize different dates by the volume of complaints received, no matter the product:

```
In [79]: plt.figure()
In [80]: for label in np.unique(model.labels_):
    ....:     if sum(model.labels_==label) > 1:
    ....:         object = df[model.labels_==label].sum(axis = 1)
    ....:         object.plot(label=label)
    ....:
In [81]: plt.legend(); \
    ....: plt.show()
```

We should get an output similar to the following:

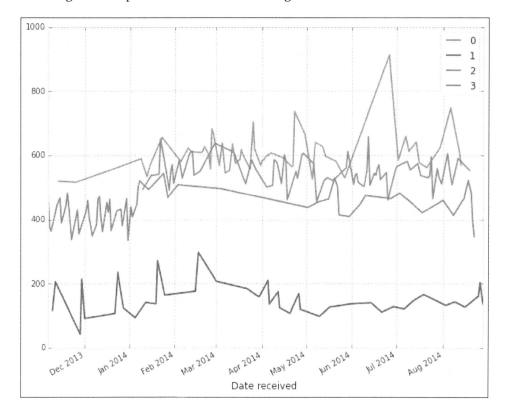

The cluster that has not been represented in this plot is, as in our previous clustering analysis, is the single date July 26, 2014, when nearly a thousand complaints were received.

In case of huge amounts of data (more than 10,000 points), we often use a variation of K-means that runs on randomly sampled subsets of the data on different iterations, to reduce computation time. This method is called **Mini-batch Kmeans**, and it has been implemented as the class `MiniBatchKMeans` in the same submodule. The quality of the clustering is slightly worse as compared to when using pure K-means, but the process is significantly faster.

Spectral clustering

By performing a low-dimension spectral embedding of the data (with different metrics) prior to a K-means, we are often able to tackle clustering when any of the previous methods fail to categorize data in a meaningful way. We have an amazingly clever implementation based on algebraic multigrid solvers in the `scikit-learn` toolkit, as the class `SpectralClustering` in the submodule `sklearn.cluster`.

> To handle the algebraic multigrid solvers, it is highly recommended to have the package `pyamg` installed. This package was developed by Nathan Bell, Luke Olson, and Jacob Schroder, from the department of computer science at the University of Illinois at Urbana-Champaign. This is not strictly necessary, but doing so will speed up our computations immensely. The package can be downloaded in several formats from `http://pyamg.org/` or installed as usual with either a `pip`, `easy_install` or `conda` command from the console.

Summary

In this chapter, we explored advanced techniques in the SciPy stack to perform inferential statistics, data mining, and machine learning. In the next chapter, we change gears completely to master mathematical imaging.

Mathematical Imaging

Mathematical Imaging is a very broad field that is concerned with the treatment of images by representing them as mathematical objects. Depending on the goals, we have four subfields:

- **Image acquisition**: The concern here is the effective representation of an object as an image. Clear examples are the digitalization of a photograph (that could be coded as a set of numerical arrays), or super-imposed information of the highest daily temperatures on a map (that could be coded as a discretization of a multivariate function). The processes of acquisition differ depending on what needs to be measured and the hardware that performs the measures. This topic is beyond the scope of this book but, if interested, some previous background can be obtained by studying the Python interface to OpenCV and any of the background libraries, such as **Python Imaging Library** (**PIL**) and the friendly PIL fork Pillow.

A nice documentation for PIL can be accessed through the `http://effbot.org/` pages at `http://effbot.org/imagingbook/pil-index.htm`. Installing the SciPy stack immediately places a copy of the latest version of PIL in our system. If needed, downloads of this library alone are available from `http://pythonware.com/products/pil/`. For information about Pillow, a good source is `http://pillow.readthedocs.org/`.

A good source of information for OpenCV can be found at `http://opencv.org/`. For a closer look at the interface to Python, I have found the tutorials at `http://docs.opencv.org/3.0-beta/doc/py_tutorials/py_tutorials.html` very useful.

Note that the installation of OpenCV for Python is not easy. My recommendation is to perform such an installation from Anaconda or any other scientific Python distribution.

- **Image compression**: This is the most technical of these subfields and requires mostly high-level libraries from NumPy, SciPy, and some extra packages. The goal is the representation of images using the minimum possible data, in a way that most (ideally all) of the relevant information is kept.

- **Image editing**: This, together with the following image analysis, is what we refer to as **Image processing**. Examples of the goals of image editing range from the restoration of damaged photographs, to the deblurring of a video sequence, or the removal of an object in an image, so that the removed area gets inpainted with coherent information. To deal with these operations, in the SciPy stack, we have the library `scipy.ndimage`, and the image processing toolkit `scikit-image`.

A good set of references and documentation for the multidimensional image-processing library `scipy.ndimage` can be found at `http://docs.scipy.org/doc/scipy/reference/tutorial/ndimage.html`, including an enlightening introduction to filters.

To explore the image processing toolkit `scikit-image`, a good initial resource is the documentation of the official page at `http://scikit-image.org/docs/stable/`. This includes a crash course on using NumPy for images.

- **Image analysis**: This is an interesting field, where we aim to obtain different pieces of information from an object represented as an image. Think of a code that can track the face of an individual in the video rendering of a large crowd, or count the number of gold atoms on a micrograph of a catalyst. For these tasks, we usually mix functions from the previous two libraries, with the ever-useful toolkit `scikit-learn` that we discussed in the previous chapter.

In our exposition, we will start with a small section on how to represent digital images within the SciPy stack. We continue with a second section on the nature of basic operations over images. The rest of the sections continue with the presentation of techniques for compression, editing, and analysis, in that order.

Most of the operations we introduce conclude with a visualization of the examples. The corresponding code is usually a trivial application of commands from `matplotlib`. These codes are usually not included, and left to the reader as exercise. Only when a specific complex layout or novel idea is introduced, we will include those codes in our presentation.

Digital images

The dictionary defines a pixel (an abbreviation of picture element) as a minute area of illumination on a display screen, one of many from which an image is composed. We therefore consider a digital image as a set of pixels, each of them defined by its location (irrespective of the kind of coordinates chosen) and the intensity of light of the corresponding image at that location.

Depending on the way we measure intensity, a digital image belongs to one of three possible types:

- Binary
- Gray-scale
- Color (with or without an alpha channel)

Binary

In a binary image there are only two possible intensities—light or dark. Such images are traditionally best implemented as simple two-dimensional Boolean arrays. True indicates a bright spot, while False measures a dark spot.

For instance, to create a binary image of size 128 x 128, with a single disk of radius 6 centered at the location (30, 100), we could issue the following:

```
In [1]: import numpy as np, matplotlib.pyplot as plt
In [2]: disk = lambda x,y: (x-30)**2 + (y-100)**2 <= 36
In [3]: image = np.fromfunction(disk, (128, 128))
In [4]: image.dtype
Out[4]: dtype('bool')
```

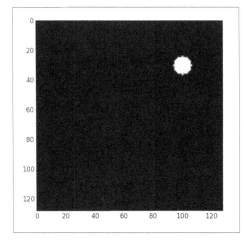

Another method to generate geometric shapes on a binary image is the set of utilities in the modules `skimage.draw` or `skimage.morphology`. For instance, the previous could have been generated as follows:

```
>>> from skimage.draw import circle
>>> image = np.zeros((128, 128)).astype('bool')
>>> image[circle(30, 100, 6)] = True
```

The module `skimage.draw` has routines to create other two-dimensional geometric shapes:

- **Lines**: `line`. There is also an anti-aliased version of a line, for gray-scale images: `line_aa`.
- **Circles**: `circle`, `circle_perimeter`. There is also an anti-aliased version of a circle perimeter, for gray-scale images: `circle_perimeter_aa`.
- **Ellipses**: `ellipse`, `ellipse_perimeter`.
- **Polygons**: `polygon`.

Gray-scale

A gray-scale image is the traditional method of representing black and white photographs. In these images, the intensity of the light is represented as different scales of gray. White indicates the brightest, and black signifies no light. The number of different scales is predetermined, and usually a dyadic number (we could choose as little as 16 scales, or as many as 256, for example). In any case, the highest value is always reserved for the brightest color (white), and the lowest for the darkest (black). A simple two-dimensional array is a good way to store this information.

The `scipy.misc` library has a test image conforming to this category. In the toolkit `skimage`, we also have a few test images with the same characteristics:

```
In [6]: from scipy.misc import lena; \
   ...: from skimage.data import coins
In [7]: lena().shape
Out[7]: (512, 512)
In [8]: lena()
Out[8]:
array([[162, 162, 162, ..., 170, 155, 128],
       [162, 162, 162, ..., 170, 155, 128],
       [162, 162, 162, ..., 170, 155, 128],
       ...,
```

```
       [ 43,   43,   50, ...,  104,  100,   98],
       [ 44,   44,   55, ...,  104,  105,  108],
       [ 44,   44,   55, ...,  104,  105,  108]])
In [9]: coins().shape
Out[9]: (303, 384)
In [10]: coins()
Out[10]:
array([[ 47, 123, 133, ...,   14,    3,   12],
       [ 93, 144, 145, ...,   12,    7,    7],
       [126, 147, 143, ...,    2,   13,    3],
       ...,
       [ 81,  79,  74, ...,    6,    4,    7],
       [ 88,  82,  74, ...,    5,    7,    8],
       [ 91,  79,  68, ...,    4,   10,    7]], dtype=uint8)
```

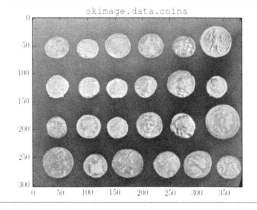

 On the left-hand side, we can see Lena, the standard (and controversial) test image scanned from the November 1972 edition of Playboy magazine. On the right-hand side, we can see Greek coins from Pompeii; this image has been downloaded from the Brooklyn Museum collection.

Color

In color images, we have many different methods to store the underlying information. The most common method, the one that also provides the easiest computational structures for creating algorithms, is the RGB color space. In this method, an image representation contains at least three layers. For each pixel, we assess the combined information of the amounts of red, green, and blue necessary to achieve the desired color and intensity at the corresponding location. The first layer indicates the intensities of underlying reds. The second and third layers indicate, respectively, the intensities of greens and blues:

```
In [12]: from skimage.data import coffee
In [13]: coffee().shape
Out[13]: (400, 600, 3)
In [14]: coffee()
Out[14]:
array([[[ 21,   13,    8],
        [ 21,   13,    9],
        [ 20,   11,    8],
        ...,
        [228, 182, 138],
        [231, 185, 142],
        [228, 184, 140]],


       ...,


       [[197, 141, 100],
        [195, 137,  99],
        [193, 138,  98],
        ...,
        [158,  73,  38],
        [144,  64,  30],
        [143,  60,  29]]], dtype=uint8)
```

This photograph, taken by Rachel Michetti, is courtesy of Pikolo Espresso Bar.

To collect the data corresponding to each of the layers, we issue simple slicing operations:

```
In [15]: plt.figure(); \
    ....: plt.subplot(131); \
    ....: plt.imshow(coffee()[:,:,0], cmap=plt.cm.Reds); \
    ....: plt.subplot(132); \
    ....: plt.imshow(coffee()[:,:,1], cmap=plt.cm.Greens); \
    ....: plt.subplot(133); \
    ....: plt.imshow(coffee()[:,:,2], cmap=plt.cm.Blues); \
    ....: plt.show()
```

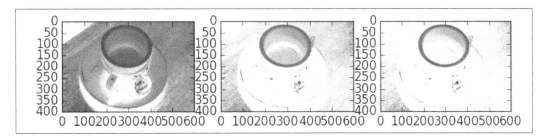

All functions in the libraries of the SciPy stack described in this chapter assume that any color image is represented in this scheme. There are other color schemes, which are designed to address other fundamental questions and properties of images. A process to transform among most of the frequent color spaces is available in the toolkit `scikit-image`, as the function `convert-colorspace` in the submodule `skimage.color`. For example, consider the **Hue-Saturation-Value (HSV)** color space. This is a cylindrical-coordinate representation of points from an RGB color space, where the angle around the central vertical axis corresponds to **hue (H)** and the distance from the axis corresponds to **saturation (S)**. The height corresponds to a third **value (V)**, the system's representation of the perceived luminance (think brightness of the underlying combination of colors) in relation to the saturation:

```
In [16]: from skimage.color import convert_colorspace
In [17]: convert_colorspace(coffee(), 'RGB', 'HSV')
Out[17]:
array([[[ 0.06410256,  0.61904762,  0.08235294],
        [ 0.05555556,  0.57142857,  0.08235294],
        [ 0.04166667,  0.6       ,  0.07843137],
        ...,
        [ 0.08148148,  0.39473684,  0.89411765],
        [ 0.08052434,  0.38528139,  0.90588235],
        [ 0.08333333,  0.38596491,  0.89411765]],


       ...,


       [[ 0.07044674,  0.49238579,  0.77254902],
        [ 0.06597222,  0.49230769,  0.76470588],
        [ 0.07017544,  0.49222798,  0.75686275],
        ...,
        [ 0.04861111,  0.75949367,  0.61960784],
        [ 0.0497076 ,  0.79166667,  0.56470588],
        [ 0.04532164,  0.7972028 ,  0.56078431]]])
```

Among the other color spaces available are the CIE XYZ method of measuring tristimulus values, or the CIE-LUB and CIE-LAB color spaces. The best resource to learn how to access and use them in the SciPy stack environment is the documentation of the module `skimage.color` in the pages of their official documentation at `http://scikit-image.org/docs/stable/api/skimage.color.html`.

It is also possible to produce a gray-scale version of any color image provided in the RGB color space, by adding the three layers with appropriate weights. In the `skimage.color` module, we have, for instance, the functions `rgb2gray` or `rgb2grey`, that employ the formula `output = 0.2125*red + 0.7154*green + 0.0721*blue`.

Alpha channels

In either gray-scale or color images, we sometimes indicate an extra layer that gives us information about the opacity of each pixel. This is referred to as the **alpha channel**. Traditionally, we incorporate this property of our images as an additional layer of an RGB—the so-called RGBA color space. In that case, images represented by this scheme have four layers instead of three:

```
In [18]: from skimage.data import horse
In [19]: horse().shape
Out[19]: (328, 400, 4)
In [20]: horse()
Out[20]:
array([[[255, 255, 255, 110],
        [255, 255, 255, 217],
        [255, 255, 255, 255],
        ...,
        [255, 255, 255, 255],
        [255, 255, 255, 217],
        [255, 255, 255, 110]],

       ...,

       [[255, 255, 255, 110],
        [255, 255, 255, 217],
        [255, 255, 255, 255],
        ...,
        [255, 255, 255, 255],
        [255, 255, 255, 217],
        [255, 255, 255, 110]]], dtype=uint8)
```

High-level operations on digital images

Before addressing the challenges of image processing and compression, it pays off to examine the sequence of basic operations that we perform on images. These operations are the building blocks of the algorithms we are to explore. By themselves, they beautifully illustrate the basic principle of mathematical imaging. An image is represented as a mathematical object and, as such, basic operations on the corresponding object can be translated to either physical operations or queries on the corresponding image.

Object measurements

In the setting of binary images, we could regard an image as a set of objects or blobs (in white) on an empty region of the plane (the background, in black). It is then possible to perform different measures on each of the objects represented:

```
In [1]: import numpy as np, matplotlib.pyplot as plt; \
   ...: from skimage.data import hubble_deep_field
In [2]: image = (hubble_deep_field()[:,:,0] > 120)
```

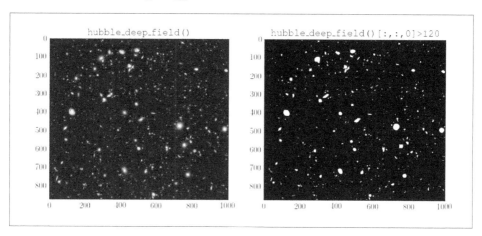

 Hubble eXtreme Deep Field: The image on the left shows the farthest view of the universe. It was captured for NASA by the Hubble Telescope and uploaded to http://hubblesite.org/. It can be freely used in the public domain.

The image on the right (a binary image) collects, as subsets of True values, a selection of the celestial objects represented in the original image. We obtained this binary image by a simple thresholding operation, where we asked for those pixels where the red intensity of the picture is greater than 120.

From this binary image, we can easily label and count the selected celestial objects, and compute some of their geometric properties. We do so by using the function `label` in the library `scipy.ndimage`:

```
In [4]: from scipy.ndimage import label
In [5]: labels, num_features = label(image); \
   ...: print "Image contains {} objects".format(num_features)
Image contains 727 objects.
```

Computing the center of mass of each object, for example, is performed with the function `center_of_mass`:

```
In [6]: from scipy.ndimage import center_of_mass
In [7]: for k in range(1,11):
   ...:       location = str(center_of_mass(image, labels, k))
   ...:       print "Object {} center of mass at {}".format(k,location)
   ...:
Object  1 center of mass at (0.0, 875.5)
Object  2 center of mass at (4.7142857142857144, 64.857142857142861)
Object  3 center of mass at (3.3999999999999999, 152.19999999999999)
Object  4 center of mass at (6.0454545454545459, 206.13636363636363)
Object  5 center of mass at (5.0, 489.5)
Object  6 center of mass at (6.5, 858.0)
Object  7 center of mass at (6.0, 586.5)
Object  8 center of mass at (7.111111111111107, 610.66666666666663)
Object  9 center of mass at (10.880000000000001, 297.45999999999998)
Object 10 center of mass at (12.800000000000001, 132.40000000000001)
```

Mathematical morphology

Again, in the setting of binary images, we have another set of interesting operations, **Mathematical morphology**. A basic morphological operation consists of probing the shape of the blobs with a common structuring element. Consider, for example, the basic operations of the erosion and dilation of shapes, using a small square as structuring element.

The erosion of an object is the set of points of said object that can be reached by the center of the structuring element, when this set moves inside of the original object. The dilation of an object, on the other hand, is the set of points covered by the structuring element, when the center of this moves inside the original object. The combination of sequences of these two operations leads to more powerful algorithms in image editing.

Also, let's observe two more advanced morphological operations: the computation of the skeleton of a shape, and the location of the medial axis of a shape (the ridges of its distance transform). These operations are also the seed of interesting processes in image analysis:

```
In [8]: from scipy.ndimage.morphology import binary_erosion; \
   ...: from scipy.ndimage.morphology import binary_dilation; \
   ...: from skimage.morphology import skeletonize, medial_axis; \
   ...: from skimage.data import horse
In [9]: image = horse()[:,:,0]==0
In [10]: # Morphology via scipy.ndimage.morphology ; \
   ....: structuring_element = np.ones((10,10)); \
   ....: erosion = binary_erosion(image, structuring_element); \
   ....: dilation = binary_dilation(image, structuring_element)
In [11]: # Morphology via skimage.morphology ; \
   ....: skeleton = skeletonize(image); \
   ....: md_axis  = medial_axis(image)
In [12]: plt.figure(); \
   ....: plt.subplot2grid((2,4), (0,0), colspan=2, rowspan=2); \
   ....: plt.imshow(image); \
   ....: plt.gray(); \
   ....: plt.title('Original Image'); \
   ....: plt.subplot2grid((2,4), (0,2)); \
   ....: plt.imshow(erosion); \
   ....: plt.title('Erosion'); \
   ....: plt.subplot2grid((2,4), (0,3)); \
   ....: plt.imshow(dilation); \
   ....: plt.title('Dilation'); \
   ....: plt.subplot2grid((2,4), (1,2)); \
   ....: plt.imshow(skeleton); \
```

```
....: plt.title('Skeleton'); \
....: plt.subplot2grid((2,4), (1,3)); \
....: plt.imshow(md_axis); \
....: plt.title('Medial Axis'); \
....: plt.show()
```

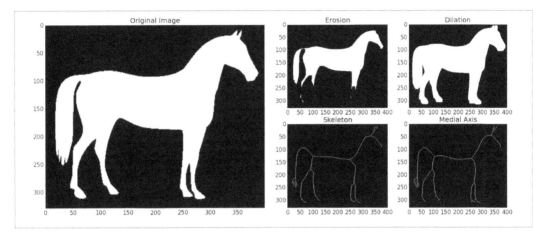

 A black and white silhouette of a horse, drawn and uploaded by Andreas Preuss into the public domain for `https://openclipart.org/`.

Smoothing filters

We can regard an image as a multivariate function. In that case, there are a few operations that compute an approximation to the original with certain good properties. A case in point is the creation of smoothed versions of images. These are the building blocks of algorithms where the presence of noise or unnecessarily complicated textures could offer confusing results.

Take for example a Gaussian filter, the convolution of a function with a Gaussian kernel with mean `mu=0` and user-defined standard `deviation sigma`:

```
In [13]: from scipy.ndimage import gaussian_filter; \
    ....: from skimage.color import rgb2gray; \
    ....: from skimage.data import coffee
In [14]: image = coffee()
In [15]: smooth_image = gaussian_filter(rgb2gray(image), sigma=2.5)
```

Note that the image on the right (the smoothed version of the original) seems blurred. For this reason, understanding the mechanisms of smoothing is also a good steppingstone with algorithms of restoration of images.

Multivariate calculus

By regarding an image as a sufficiently smooth intensity function now (with or without the help of smoothing filters) many operations are available in terms of multivariate calculus techniques. For example, the Prewitt and Sobel operators compute approximations to the norm of the gradient of said function. The corresponding values at each location assess the probability of having an edge, and are therefore used in the construction of reliable feature-detection algorithms:

```
In [17]: from scipy.ndimage import prewitt
In [18]: gradient_approx = prewitt(smooth_image)
```

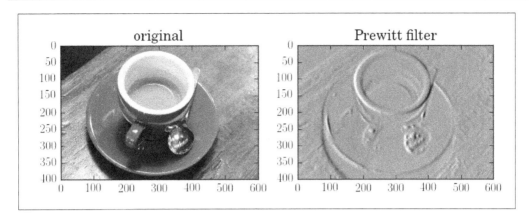

Due to the properties of the original image and its smoothed version, in this case, the absolute value of the magnitudes of the gradient (right) ranges from 0 (black) to 0.62255043871863613 (white). Brighter areas thus indicate the location of possible edges, while darker areas imply the location of flatter regions.

The sum of second-derivatives (the Laplacian operator) is also used in algorithms of feature detection or motion estimation:

```
In [20]: from scipy.ndimage import laplace
In [21]: laplace_approx = laplace(smooth_image)
```

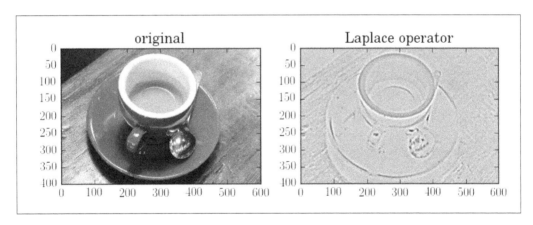

In this case, the combination of the information of the Laplacian with the gradient gives clues to the location of local extrema and the geometry of the objects represented.

The Hessian matrix (the second order partial derivatives of a scalar-valued function) is used to describe the local curvature of the image. It is a useful component in the process of the detection of blobs. We have an implementation of this operator in the module `skimage.feature`, as the routine `hessian_matrix`. For an approximation to the determinant of the Hessian, we have the routine `hessian_matrix_det` in the same module.

Statistical filters

Treating an image as a multidimensional signal, we have many filtering operations of a statistical nature available.

For instance, maximum, minimum, median, or percentile filters can be computed using the functions `maximum_filter`, `minimum_filter`, `median_filter` or `percentile_filter` in `scipy.ndimage`, respectively. These filters respectively compute, for each pixel on the image and a given footprint, the maximum, minimum, median, or requested percentile of the image on the footprint centered at the pixel.

In the following example, we compute the 80th percentile using a 10 x 10 square as a footprint:

```
In [23]: from scipy.ndimage import percentile_filter
In [24]: prctl_image = percentile_filter(image[:,:,0],
    ....:                                percentile=-20, size=10)
```

More relevant filters from this category can be found in the submodule `skimage.filters.rank`.

Fourier analysis

By regarding an image as a multivariate function again, we can perform the Fourier analysis on it. Applications of the Fourier transform, and discrete cosine transform, are mainly aimed at filtering, extracting information from the images, and compression:

```
In [25]: from scipy.fftpack import fft2, ifft2, fftshift; \
   ....: from skimage.data import text
In [26]: image = text()
In [27]: frequency = fftshift(fft2(image))
```

The frequency of a function is, in general, a complex valued function. To visualize it, we will present the module and angle of each output value. For a better interpretation, we usually enhance visually the result by applying a logarithmic correction over the module of the frequency:

```
In [28]: plt.figure(); \
   ....: ax1 = plt.subplot2grid((2,2), (0,0), colspan=2); \
   ....: plt.imshow(image)
Out[28]: <matplotlib.image.AxesImage at 0x11deb1650>
In [29]: module = np.absolute(frequency); \
   ....: angles = np.angle(frequency)
In [30]: from skimage.exposure import adjust_log
In [31]: ax2 = plt.subplot2grid((2,2), (1,0)); \
   ....: plt.imshow(adjust_log(module)); \
   ....: ax3 = plt.subplot2grid((2,2), (1,1)); \
   ....: plt.imshow(angles); \
   ....: plt.show()
```

Text is an image downloaded from Wikipedia and released to the public domain. It can be found at `https://en.wikipedia.org/wiki/File:Corner.png`.

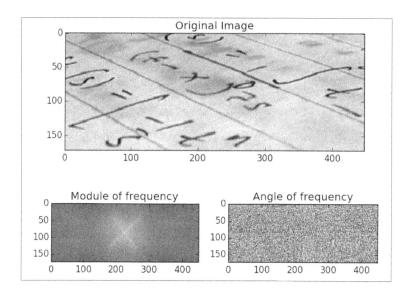

Notice what happens if we disregard part of the information from the frequency of the image — say, about 25 percent of it, and then perform an inversion:

```
In [32]: frequency.shape
Out[32]: (172, 448)
In [33]: smaller_frequency = frequency[:,448/2-172/2:448/2+172/2]
In [34]: new_image = ifft2(smaller_frequency); \
    ....: new_image = np.absolute(new_image)
```

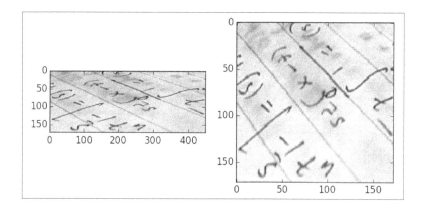

Although we are missing a fourth of the original frequency, the inversion gives us an image with exactly the same information as the original. What have we actually lost by disregarding that portion of the lower frequencies? The answer to this question leads to interesting algorithms of reconstruction, compression, and analysis.

Wavelet decompositions

We can perform wavelet decompositions using the package PyWavelets, written by Tariq Rashid.

It can be downloaded from https://pypi.python.org/pypi/ PyWavelets. Posterior installation can be performed following instructions from those pages. For some architecture, the installation might be tricky. We recommend that, in that cases you work from within a scientific Python distribution, such as Anaconda. For instance, we could search for the package using the binstar/conda commands:

```
% binstar search -t conda pywavelets
% conda install -c conda.binstar.org/dgursoy pywavelets
```

Note that there are many different families of wavelets implemented in this library:

```
In [36]: import pywt
In [37]: pywt.families()
Out[37]: ['haar', 'db', 'sym', 'coif', 'bior', 'rbio', 'dmey']
In [38]: print pywt.wavelist()
['bior1.1', 'bior1.3', 'bior1.5', 'bior2.2', 'bior2.4', 'bior2.6',
 'bior2.8', 'bior3.1', 'bior3.3', 'bior3.5', 'bior3.7', 'bior3.9',
 'bior4.4', 'bior5.5', 'bior6.8', 'coif1', 'coif2', 'coif3', 'coif4',
 'coif5', 'db1', 'db2', 'db3', 'db4', 'db5', 'db6', 'db7', 'db8',
 'db9', 'db10', 'db11', 'db12', 'db13', 'db14', 'db15', 'db16',
 'db17', 'db18', 'db19', 'db20', 'dmey', 'haar', 'rbio1.1','rbio1.3',
 'rbio1.5', 'rbio2.2', 'rbio2.4', 'rbio2.6', 'rbio2.8', 'rbio3.1',
 'rbio3.3', 'rbio3.5', 'rbio3.7', 'rbio3.9', 'rbio4.4', 'rbio5.5',
 'rbio6.8', 'sym2', 'sym3', 'sym4', 'sym5', 'sym6', 'sym7', 'sym8',
 'sym9', 'sym10', 'sym11', 'sym12', 'sym13', 'sym14', 'sym15',
 'sym16', 'sym17', 'sym18', 'sym19', 'sym20']
```

Let's see how to compute a representation of the image `skimage.data.camera` using a `haar` wavelet. Since the original image has a side length of `512=2^9`, we are going to need nine levels in the computation of the wavelet coefficients:

```
In [39]: from skimage.data import camera
In [40]: levels = int(np.floor(np.log2(camera().shape).max())); \
    ....: print "We need {} levels".format(levels)
We need 9 levels
In [41]: wavelet = pywt.Wavelet('haar')
In [42]: wavelet_coeffs = pywt.wavedec2(camera(), wavelet,
    ....:                               level=levels)
```

The object `wavelet_coeffs` is a tuple with ten entries, the first one is the approximation at the highest level 0. This is always one single coefficient. The second entry in `wavelet_coeffs` is a 3-tuple containing the three different details (horizontal, vertical, and diagonal) at level 1. Each consecutive entry is another 3-tuple containing the three different details at higher levels (`n = 2, 3, 4, 5, 6, 7, 8, 9`).

Note the number of coefficients for each level:

```
In [43]: for index, level in enumerate(wavelet_coeffs):
    ....:     if index > 0:
    ....:         value = level[0].size + level[1].size + level[2].size
    ....:         print "Level {}: {}".format(index, value)
    ....:     else:
    ....:         print "Level 0: 1"
    ....:
Level 0: 1
Level 1: 3
Level 2: 12
Level 3: 48
Level 4: 192
Level 5: 768
Level 6: 3072
Level 7: 12288
Level 8: 49152
Level 9: 196608
```

Image compression

The purpose of compression is the representation of images by methods that require less units of information (for example, bytes) than the mere storage of each pixel in arrays.

For instance, recall the binary image we constructed in the first section; that is a 128 x 128 image represented by 16,384 bits (True/False), where all but 113 of those bits are False. There surely must be more efficient ways to store this information in a way that require less than 16,384 bits. We could very well do so by simply providing the size of the canvas (two bytes), the location of the center of the disk (two more bytes), and the value of its radius (another byte). We now have a new representation using only 40 bits (assuming each byte consists of 8 bits). We refer to such exact representation as a lossless compression.

Another possible way to compress an image is the process of turning a color image into its black and white representation, for example. We performed this operation on the image skimage.data.coffee, turning an object of size 3 x 400 x 600 (720,000 bytes) into an object of size 400 x 600 (240,000 bytes). Although, in the process, we lost the ability to see its color. This kind of operation is appropriately called a lossy compression.

In the following pages, we are going to explore several settings for image compression from a mathematical point of view. We will also develop efficient code to perform these operations from within the SciPy stack. We are not concerned with the creation of code to read or save these compressed images to file; for that, we already have reliable utilities in the Python Imaging Library, that have also been imported to different functions in the modules scipy.misc, scipy.ndimage, and the toolkit scikit-image. If we wish to compress and store a numpy array A representing a black-and-white photography as different file types, we simply issue something along these lines:

```
In [1]: import numpy as np; \
   ...: from scipy.misc import lena, imsave
In [2]: A = lena()
In [3]: imsave("my_image.png", A); \
   ...: imsave("my_image.tiff", A); \
   ...: imsave("my_image.pcx", A); \
   ...: imsave("my_image.jpg", A); \
   ...: imsave("my_image.gif", A)
```

A quick visualization of the contents of the folder in which we are working shows the sizes of the files created. For instance, under a *NIX system, we could issue the following command:

```
% ls -nh my_image.*
-rw-r--r--  1 501  20    257K May 29 08:16 my_image.bmp
-rw-r--r--  1 501  20     35K May 29 08:16 my_image.jpg
-rw-r--r--  1 501  20    273K May 29 08:15 my_image.pcx
-rw-r--r--  1 501  20    256K May 29 08:16 my_image.tiff
```

Note the different sizes of the resulting files. The lossless formats PCX, BMP, and TIFF offer similar compression rates (273K, 257K, and 256 K, respectively). On the other hand, the JPEG lossy format offers an obvious improvement (35 K).

Lossless compression

Some of the most common lossless compression schemes used for image processing are as follows:

- **Run-length encoding**: This method is used in PCX, BMP, TGA, and TIFF file types, when the original image can be regarded as palette-based bitmapped, for example, a cartoon or a computer icon.

- **Lempel-Ziv-Welch (LZW)**: This is used by default in the GIF image format.

- **Deflation**: This is very powerful and reliable. It is the method used for PNG image files. It is also the compression method employed to create ZIP files.

- **Chain code**: This is the preferred method to encode binary images, especially if these contain a small number of large blobs.

Let's examine, for instance, how run-length encoding works in a suitable example. Consider the checkerboard image `skimage.data.checkerboard`. We receive it as a 200 x 200 array of integer values and, in this way, it requires 40,000 bytes of storage. Note, it can be regarded as a palette-based bitmap with only two colors. We start by transforming each zero value to a B, and each 255 to a W:

```
In [5]: from skimage.data import checkerboard
In [6]: def color(value):
   ...:       if value==0: return 'B'
   ...:       else: return 'W'
   ...:
In [7]: image = np.vectorize(color)(checkerboard()); \
   ...: print image
[['W' 'W' 'W' ..., 'B' 'B' 'B']
```

```
 ['W' 'W' 'W' ..., 'B' 'B' 'B']
 ['W' 'W' 'W' ..., 'B' 'B' 'B']
 ...,
 ['B' 'B' 'B' ..., 'W' 'W' 'W']
 ['B' 'B' 'B' ..., 'W' 'W' 'W']
 ['B' 'B' 'B' ..., 'W' 'W' 'W']]
```

Next, we create a function that encodes both lists and strings of characters, producing instead, a string composed of patterns of the form "single character plus count":

```
In [7]: from itertools import groupby
In [8]: def runlength(string):
   ...:     groups = [k + str(sum(1 for _ in g)) for k,g in
   ...:               groupby(string)]
   ...:     return ''.join(groups)
   ...:
```

Notice what happens when we rewrite the image as a flattened string containing its colors and encode it in this fashion:

```
In [9]: coded_image = runlength(image.flatten().tolist())
In [10]: print coded_image
W26B23W27B23W27B23W27B24W26B23W27B23W27B23W27B24W26B23W27B23W27B23W27
B24W26B23W27B23W27B23W27B24W26B23W27B23W27B23W27B24W26B23W27B23W27B23
W27B24W26B23W27B23W27B23W27B24W26B23W27B23W27B23W27B24W26B23W27B23W27
B23W27B24W26B23W27B23W27B23W27B24W26B23W27B23W27B23W27B24W26B23W27B23
W27B23W27B24W26B23W27B23W27B23W27B24W26B23W27B23W27B23W27B24W26B23W27
B23W27B23W27B24W26B23W27B23W27B23W27B24W26B23W27B23W27B23W27B24W26B23
W27B23W27B23W27B24W26B23W27B23W27B23W27B24W26B23W27B23W27B23W27B24W26
B23W27B23W27B23W27B24W26B23W27B23W27B23W27B24W26B
...
26B24W27B23W27B23W27B23W26B24W27B23W27B23W27B23W26B24W27B23W27B23W27B
23W26B24W27B23W27B23W27B23W26B24W27B23W27B23W27B23W26B24W27B23W27B23W
27B23W26B24W27B23W27B23W27B23W26B24W27B23W27B23W27B23W26B24W27B23W27B
23W27B23W26B24W27B23W27B23W27B23W26B24W27B23W27B23W27B23W26B24W27B23W
27B23W27B23W26B24W27B23W27B23W27B23W26B24W27B23W27B23W27B23W26B24W27B
23W27B23W27B23W26B24W27B23W27B23W27B23W26B24W27B23W27B23W27B23W26B24W
27B23W27B23W27B23W26B24W27B23W27B23W27B23W26B24W27B23W27B23W27B23W26
In [11]: len(coded_image)
Out[11]: 4474
```

We have reduced its size to a mere 4,474 bytes. Now, how would you decode this information back to an image? Imagine you are provided with this string, the additional information of the size of the image (200 x 200), and the palette information (B for black, and W for white).

Another nice exercise is to find descriptions of the other mentioned lossless compression methods, and write Python codes for their corresponding encoder and decoder.

Lossy compression

Among the many possible schemes of lossy compression, we are going to focus on the method of transform coding. The file type JPEG, for instance, is based on the discrete cosine transform.

In any of these cases, the process is similar. We assume that the image is a function. Its visualization can be regarded as a representation of its graph, and as such this is a spatial operation. Instead, we can compute a transform of the image (say, Fourier, discrete cosine, or Wavelet). The image is now represented by a collection of values: The coefficients of the function in the corresponding transform. Now, compression occurs when we disregard a large quantity of those coefficients and reconstruct the function with the corresponding inverse transform.

We have already observed the behavior of the reconstruction of an image after disregarding 25 percent of its lower frequencies, when addressing the Fourier analysis techniques in the previous section. This is not the only way to disregard coefficients. For instance, we can instead collect coefficients with a large enough absolute value. Let's examine the result of performing that operation on the same image, this time using the discrete cosine transform:

```
In [12]: from skimage.data import text; \
   ....: from scipy.fftpack import dct, idct
In [13]: image = text().astype('float')
In [14]: image_DCT = dct(image)
```

Let's disregard the values that are less than or equal to 1000 in absolute value. Note that there are more than 256,317 such coefficients (almost 98 percent of the original data):

```
In [15]: mask = np.absolute(image_DCT)>1000
In [16]: compressed = idct(image_DCT * mask)
```

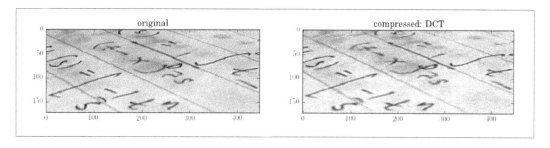

In spite of having thrown away most of the coefficients, the reconstruction is very faithful. There are obvious artifacts, but these are not too distracting.

We can perform a similar operation using the wavelet transform. A naive way to perform compression in this setting could be to disregard whole levels of coefficients, and then reconstruct. In the example that we covered in the previous sections (a Haar wavelet representation of the image `skimage.data.camera` with nine levels of coefficients), if we eliminate the last two levels, we are throwing away 245,760 coefficients (almost 94 percent of the original information). Observe the quality of the reconstruction:

```
In [18]: import pywt; \
   ....: from skimage.data import camera
In [19]: levels = int(np.floor(np.log2(camera().shape).max())))
In [20]: wavelet = pywt.Wavelet('haar')
In [21]: wavelet_coeffs = pywt.wavedec2(camera(), wavelet,
   ....:                                level=levels)
In [22]: compressed = pywt.waverec2(wavelet_coeffs[:8], wavelet)
```

Similar to transform coding is the method of compression by singular value decomposition. In this case, we regard an image as a matrix. We represent it by means of its singular values. Compression in this setting occurs when we disregard a large quantity of the smaller singular values, and then reconstruct. For an example of this technique, read *Chapter 3, SciPy for Linear Algebra* of the book *Learning SciPy for Numerical and Scientific Computing, Second Edition*.

Image editing

The purpose of editing is the alteration of digital images, usually to improve its properties or to turn them into an intermediate step for further processing.

Let's examine different methods of editing:

- Transformations of the domain
- Intensity adjustment
- Image restoration
- Image inpainting

Transformations of the domain

In this setting, we address transformations to images by first changing the location of pixels: rotations, compressions, stretching, swirls, cropping, perspective control, and so on. Once the transformation to the pixels in the domain of the original is performed, we observe the size of the output. If there are more pixels in this image than in the original, the extra locations are filled with numerical values obtained by interpolating the given data. We do have some control over the kind of interpolation performed, of course. To better illustrate these techniques, we will pair an actual image (say, Lena) with a representation of its domain as a checkerboard:

```
In [1]: import numpy as np, matplotlib.pyplot as plt
In [2]: from scipy.misc import lena; \
   ...: from skimage.data import checkerboard
In [3]: image = lena().astype('float')
   ...: domain = checkerboard()
In [4]: print image.shape, domain.shape
Out[4]: (512, 512) (200, 200)
```

Rescale and resize

Before we proceed with the pairing of image and domain, we have to make sure that they both have the same size. One quick fix is to rescale both objects, or simply resize one of the images to match the size of the other. Let's go with the first choice, so that we can illustrate the usage of the two functions available for this task in the module `skimage.transform` to resize and rescale:

```
In [5]: from skimage.transform import rescale, resize
In [6]: domain = rescale(domain, scale=1024./200); \
   ...: image  = resize(image, output_shape=(1024, 1024), order=3)
```

Observe how, in the resizing operation, we requested a bicubic interpolation.

Swirl

To perform a swirl, we call the function `swirl` from the module `skimage.transform`:

In all the examples of this section, we will present the results visually after performing the requested computations. In all cases, the syntax of the call to offer the images is the same. For a given operation mapping, we issue the command `display(mapping, image, domain)` where the routine `display` is defined as follows:

```
def display(mapping, image, domain):
    plt.figure()
    plt.subplot(121)
    plt.imshow(mapping(image))
    plt.gray()
    plt.subplot(122)
    plt.imshow(mapping(domain))
    plt.show()
```

For the sake of brevity, we will not include this command in the following code, but assume it is called every time:

```
In [7]: from skimage.transform import swirl
In [8]: def mapping(img):
   ...:         return swirl(img, strength=6, radius=512)
   ...:
```

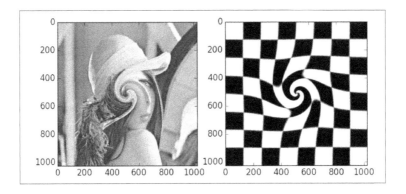

Geometric transformations

A simple rotation around any location (no matter whether inside or outside the domain of the image) can be achieved with the function rotate from either module `scipy.ndimage` or `skimage.transform`. They are basically the same under the hood, but the syntax of the function in the `scikit-image` toolkit is more user friendly:

```
In [10]: from skimage.transform import rotate
In [11]: def mapping(img):
   ....:         return rotate(img, angle=30, resize=True, center=None)
   ....:
```

This gives a counter-clockwise rotation of 30 degrees (`angle=30`) around the center of the image (`center=None`). The size of the output image is expanded to guarantee that all the original pixels are present in the output (`resize=True`):

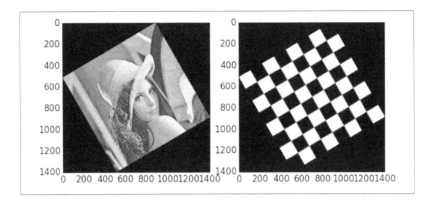

Rotations are a special case of what we call an affine transformation—a combination of rotation with scales (one for each dimension), shear, and translation. Affine transformations are in turn a special case of a homography (a projective transformation). Rather than learning a different set of functions, one for each kind of geometric transformation, the library `skimage.transform` allows a very comfortable setting. There is a common function (`warp`) that gets called with the requested geometric transformation and performs the computations. Each suitable geometric transformation is previously initialized with an appropriate Python class. For example, to perform an affine transformation with a counter-clockwise rotation angle of 30 degrees about the point with coordinates (512, -2048), and scale factors of 2 and 3 units, respectively, for the x and y coordinates, we issue the following command:

```
In [13]: from skimage.transform import warp, AffineTransform
In [14]: operation = AffineTransform(scale=(2,3), rotation=np.pi/6, \
   ....:                              translation = (512, -2048))
In [15]: def mapping(img):
   ....:      return warp(img, operation)
   ....:
```

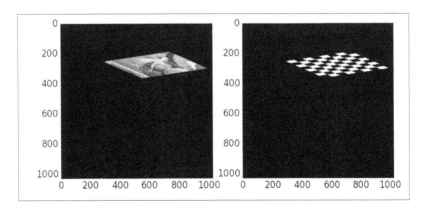

Observe how all the lines in the transformed checkerboard are either parallel or perpendicular—affine transformations preserve angles.

The following illustrates the effect of a homography:

```
In [17]: from skimage.transform import ProjectiveTransform
In [18]: generator = np.matrix('1,0,10; 0,1,20; -0.0007,0.0005,1'); \
    ....: homography = ProjectiveTransform(matrix=generator); \
    ....: mapping = lambda img: warp(img, homography)
```

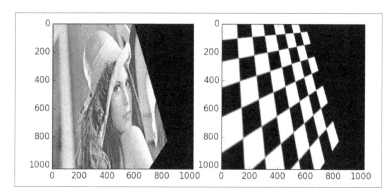

Observe how, unlike in the case of an affine transformation, the lines cease to be all parallel or perpendicular. All vertical lines are now incident at a point. All horizontal lines are also incident at a different point.

The real usefulness of homographies arises, for example, when we need to perform perspective control. For instance, the image `skimage.data.text` is clearly slanted. By choosing the four corners of what we believe is a perfect rectangle (we estimate such a rectangle by visual inspection), we can compute a homography that transforms the given image into one that is devoid of any perspective. The Python classes representing geometric transformations allow us to perform this estimation very easily, as the following example shows:

```
In [20]: from skimage.data import text
In [21]: text().shape
Out[21]: (172, 448)
In [22]: source = np.array(((155, 15),  (65, 40),
    ....:                   (260, 130), (360, 95),
    ....:                   (155, 15)))
In [23]: mapping = ProjectiveTransform()
```

Let's estimate the homography that transforms the given set of points into a perfect rectangle of the size 48 x 256 centered in an output image of size 512 x 512. The choice of size of the output image is determined by the length of the diagonal of the original image (about 479 pixels). This way, after the homography is computed, the output is likely to contain all the pixels from the original:

 Observe that we have included one of the vertices in the source twice. This is not strictly necessary for the following computations, but will make the visualization of rectangles much easier to code. We use the same trick for the target rectangle.

```
In [24]: target = np.array(((256-128, 256-24), (256-128, 256+24),
    ....:                    (256+128, 256+24), (256+128, 256-24),
    ....:                    (256-128, 256-24)))
In [25]: mapping.estimate(target, source)
Out[25]: True
In [26]: plt.figure(); \
    ....: plt.subplot(121); \
    ....: plt.imshow(text()); \
    ....: plt.gray(); \
    ....: plt.plot(source[:,0], source[:,1],'-', lw=1, color='red'); \
    ....: plt.xlim(0, 448); \
    ....: plt.ylim(172, 0); \
    ....: plt.subplot(122); \
    ....: plt.imshow(warp(text(), mapping,output_shape=(512, 512))); \
    ....: plt.plot(target[:,0], target[:,1],'-', lw=1, color='red'); \
    ....: plt.xlim(0, 512); \
    ....: plt.ylim(512, 0); \
    ....: plt.show()
```

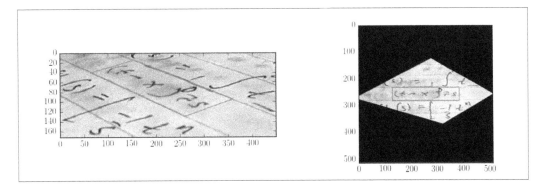

Other more involved geometric operations are needed, for example, to fix vignetting—and some of the other kinds of distortions produced by photographic lenses. Traditionally, once we acquire an image we assume that all these distortions are present. By knowing the technical specifications of the equipment used to take the photographs, we can automatically rectify these defects. With this purpose in mind, in the SciPy stack we have access to the `lensfun` library (`http://lensfun.sourceforge.net/`) through the package `lensfunpy` (`https://pypi.python.org/pypi/lensfunpy`).

For examples of usage and documentation, an excellent resource is the API reference of `lensfunpy` at `http://pythonhosted.org/lensfunpy/api/`.

Intensity adjustment

In this category, we have operations that only modify the intensity of an image obeying a global formula. All these operations can therefore be easily coded by using purely NumPy operations, by creating vectorized functions adapting the requested formulas.

The applications of these operations can be explained in terms of exposure in black and white photography, for example. For this reason, all the examples in this section are applied on gray-scale images.

We have mainly three approaches to enhancing images by working on its intensity:

- Histogram equalization
- Intensity clipping/resizing
- Contrast adjustment

Histogram equalization

The starting point in this setting is always the concept of intensity histogram (or more precisely, the histogram of pixel intensity values)—a function that indicates the number of pixels in an image at each different intensity value found in that image.

For instance, for the original version of Lena, we could issue the following command:

```
In [27]: plt.figure(); \
    ....: plt.hist(lena().flatten(), 256); \
    ....: plt.show()
```

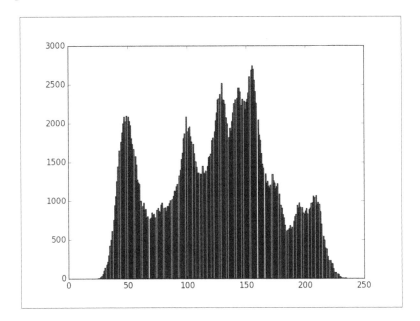

The operations of histogram equalization improve the contrast of images by modifying the histogram in a way so that most of the relevant intensities have the same impact. We can accomplish this enhancement by calling, from the module `skimage.exposure`, any of the functions `equalize_hist` (pure histogram equalization) or `equalize_adaphist` (**contrast limited adaptive histogram equalization (CLAHE)**).

Note the obvious improvement after the application of histogram equalization to the image `skimage.data.moon`:

> In the following examples, we include the corresponding histogram below all relevant images for comparison. A suitable code to perform this visualization could be as follows:
>
> ```
> def display(image, transform, bins):
>     target = transform(image)
>     plt.figure()
>     plt.subplot(221)
>     plt.imshow(image)
>     plt.gray()
>     plt.subplot(222)
>     plt.imshow(target)
>     plt.subplot(223)
>     plt.hist(image.flatten(), bins)
>     plt.subplot(224)
>     plt.hist(target.flatten(), bins)
>     plt.show()
> ```

```
In [28]: from skimage.exposure import equalize_hist; \
   ....: from skimage.data import moon
In [29]: display(moon(), equalize_hist, 256)
```

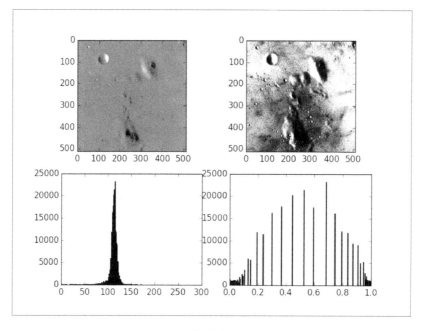

Intensity clipping/resizing

A peak at the histogram indicates the presence of a particular intensity that is remarkably more predominant than its neighboring ones. If we desire to isolate intensities around a peak, we can do so using purely NumPy masking/clipping operations on the original image. If storing the result is not needed, we can request a quick visualization of the result by employing the command `clim` in the library `matplotlib.pyplot`. For instance, to isolate intensities around the highest peak of Lena (roughly, these are between 150 and 160), we could issue the following command:

```
In [30]: plt.figure(); \
   ....: plt.imshow(lena()); \
   ....: plt.clim(vmin=150, vmax=160); \
   ....: plt.show()
```

Note how this operation, in spite of having reduced the representative range of intensities from 256 to 10, offers us a new image that has sufficient information to recreate the original one. Naturally, we can regard this operation also as a lossy compression method:

Contrast enhancement

An obvious drawback of clipping intensities is the loss of perceived lightness contrast. To overcome this loss, it is preferable to employ formulas that do not reduce the size of the range. Among the many available formulas conforming to this mathematical property, the most successful ones are those that replicate an optical property of the acquisition method. We explore the following three cases:

- **Gamma correction**: Human vision follows a power function, with greater sensitivity to relative differences between darker tones than between lighter ones. Each original image, as captured by the acquisition device, might allocate too many bits or too much bandwidth to highlight that humans cannot actually differentiate. Similarly, too few bits/bandwidth could be allocated to the darker areas of the image. By manipulation of this power function, we are capable of addressing the correct amount of bits and bandwidth.

- **Sigmoid correction**: Independently of the amount of bits and bandwidth, it is often desirable to maintain the perceived lightness contrast. Sigmoidal remapping functions were then designed based on an empirical contrast enhancement model developed from the results of psychophysical adjustment experiments.

- **Logarithmic correction**: This is a purely mathematical process designed to spread the range of naturally low-contrast images by transforming to a logarithmic range.

To perform gamma correction on images, we could employ the function `adjust_gamma` in the module `skimage.exposure`. The equivalent mathematical operation is the power-law relationship `output = gain * input^gamma`. Observe the improvement in the definition of the brighter areas of a stained micrograph of colonic glands, when we choose the exponent `gamma=2.5` and no gain (`gain=1.0`):

```
In [31]: from skimage.exposure import adjust_gamma; \
   ....: from skimage.color import rgb2gray; \
   ....: from skimage.data import immunohistochemistry
In [32]: image = rgb2gray(immunohistochemistry())
In [33]: correction = lambda img: adjust_gamma(img, gamma=2.5,
   ....:                                        gain=1.)
```

Note the huge improvement in contrast in the lower-right section of the micrograph, allowing a better description and differentiation of the observed objects:

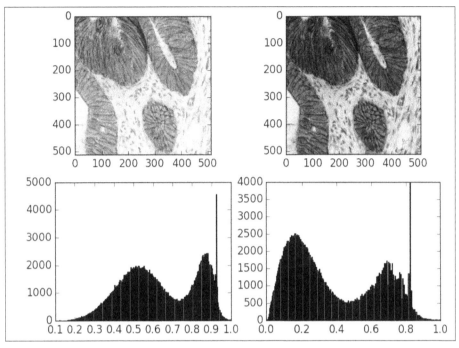

Immunohistochemical staining with hematoxylin counterstaining. This image was acquired at the Center for Microscopy And Molecular Imaging (CMMI).

To perform sigmoid correction with given `gain` and `cutoff` coefficients, according to the formula `output = 1/(1 + exp*(gain*(cutoff - input)))`, we employ the function `adjust_sigmoid` in `skimage.exposure`. For example, with `gain=10.0` and `cutoff=0.5` (the default values), we obtain the following enhancement:

```
In [35]: from skimage.exposure import adjust_sigmoid
In [36]: display(image[:256, :256], adjust_sigmoid, 256)
```

Note the improvement in the definition of the walls of cells in the enhanced image:

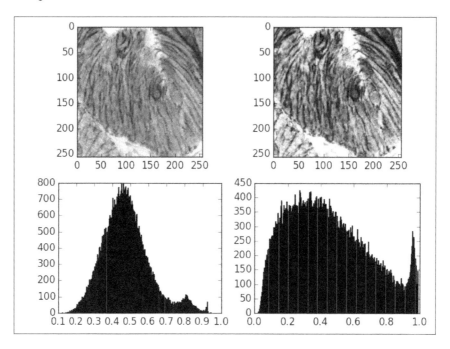

We have already explored logarithmic correction in the previous section, when enhancing the visualization of the frequency of an image. This is equivalent to applying a vectorized version of np.log1p to the intensities. The corresponding function in the scikit-image toolkit is adjust_log in the sub module exposure.

Image restoration

In this category of image editing, the purpose is to repair the damage incurred by post or preprocessing of the image, or even the removal of distortion produced by the acquisition device. We explore two classic situations:

* Noise reduction
* Sharpening and blurring

Noise reduction

In mathematical imaging, noise is by definition a random variation of the intensity (or the color) produced by the acquisition device. Among all the possible types of noise, we acknowledge four key cases:

- **Gaussian noise**: We add to each pixel a value obtained from a random variable with normal distribution and a fixed mean. We usually allow the same variance on each pixel of the image, but it is feasible to change the variance depending on the location.

- **Poisson noise**: To each pixel, we add a value obtained from a random variable with Poisson distribution.

- **Salt and pepper**: This is a replacement noise, where some pixels are substituted by zeros (black or pepper), and some pixels are substituted by ones (white or salt).

- **Speckle**: This is a multiplicative kind of noise, where each pixel gets modified by the formula `output = input + n * input`. The value of the modifier `n` is a value obtained from a random variable with uniform distribution of fixed mean and variance.

To emulate all these kinds of noise, we employ the utility `random_noise` from the module `skimage.util`. Let's illustrate the possibilities in a common image:

```
In [37]: from skimage.data import camera; \
   ....: from skimage.util import random_noise
In [38]: gaussian_noise = random_noise(camera(), 'gaussian',
   ....:                               var=0.025); \
   ....: poisson_noise = random_noise(camera(), 'poisson'); \
   ....: saltpepper_noise = random_noise(camera(), 's&p',
   ....:                                 salt_vs_pepper=0.45); \
   ....: speckle_noise = random_noise(camera(), 'speckle', var=0.02)
In [39]: variance_generator = lambda i,j: 0.25*(i+j)/1022. + 0.001; \
   ....: variances = np.fromfunction(variance_generator,(512,512)); \
   ....: lclvr_noise = random_noise(camera(), 'localvar',
   ....:                            local_vars=variances)
```

In the last example, we have created a function that assigns a variance between `0.001` and `0.026` depending on the distance to the upper-corner of an image. When we visualize the corresponding noisy version of `skimage.data.camera`, we see that the level of degradation gets stronger as we get closer to the lower-right corner of the picture.

The following is an example of the visualization of the corresponding noisy images:

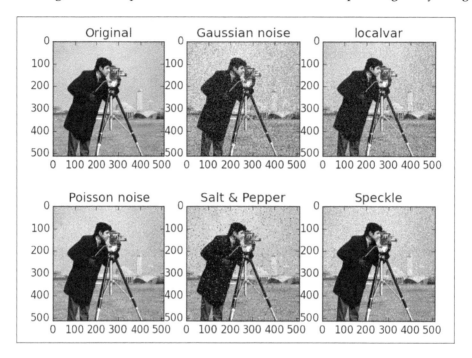

The purpose of noise reduction is to remove as much of this unwanted signal, so we obtain an image as close to the original as possible. The trick, of course, is to do so without any previous knowledge of the properties of the noise.

The most basic methods of denoising are the application of either a Gaussian or a median filter. We explored them both in the previous section. The former was presented as a smoothing filter (`gaussian_filter`), and the latter was discussed when we explored statistical filters (`median_filter`). They both offer decent noise removal, but they introduce unwanted artifacts as well. For example, the Gaussian filter does not preserve edges in images. The application of any of these methods is also not recommended if preserving texture information is needed.

We have a few more advanced methods in the module `skimage.restoration`, able to tailor denoising to images with specific properties:

- `denoise_bilateral`: This is the bilateral filter. It is useful when preserving edges is important.

- `denoise_tv_bregman, denoise_tv_chambolle`: We will use this if we require a denoised image with small total variation.

- `nl_means_denoising`: The so-called non-local means denoising. This method ensures the best results for denoising areas of the image presenting texture.

- `wiener, unsupervised_wiener`: This is the Wiener-Hunt deconvolution. It is useful when we have knowledge of the point-spread function at acquisition time.

Let us show by example, the performance of one of these methods on some of the noisy images we computed earlier:

```
In [40]: from skimage.restoration import nl_means_denoising as dnoise
In [41]: images = [gaussian_noise, poisson_noise,
   ....:           saltpepper_noise, speckle_noise]; \
   ....: names  = ['Gaussian', 'Poisson', 'Salt & Pepper', 'Speckle']
In [42]: plt.figure()
Out[42]: <matplotlib.figure.Figure at 0x118301490>
In [43]: for index, image in enumerate(images):
   ....:     output = dnoise(image, patch_size=5, patch_distance=7)
   ....:     plt.subplot(2, 4, index+1)
   ....:     plt.imshow(image)
   ....:     plt.gray()
   ....:     plt.title(names[index])
   ....:     plt.subplot(2, 4, index+5)
   ....:     plt.imshow(output)
   ....:
In [44]: plt.show()
```

Under each noisy image, we have presented the corresponding result after employing non-local means denoising.

It is also possible to perform denoising by `thresholding` coefficients, provided we represent images with a transform. For example, to do a soft thresholding, employing Biorthonormal 2.8 wavelets, we would use the package `PyWavelets`:

```
In [45]: import pywt
In [46]: def dnoise(image, wavelet, noise_var):
   ....:     levels = int(np.floor(np.log2(image.shape[0])))
   ....:     coeffs = pywt.wavedec2(image, wavelet, level=levels)
   ....:     value = noise_var * np.sqrt(2 * np.log2(image.size))
   ....:     threshold = lambda x: pywt.thresholding.soft(x, value)
   ....:     coeffs = map(threshold, coeffs)
   ....:     return pywt.waverec2(coeffs, wavelet)
   ....:
In [47]: plt.figure()
Out[47]: <matplotlib.figure.Figure at 0x10e5ed790>
In [48]: for index, image in enumerate(images):
   ....:     output = dnoise(image, pywt.Wavelet('bior2.8'),
```

```
    ....:                          noise_var=0.02)
    ....:        plt.subplot(2, 4, index+1)
    ....:        plt.imshow(image)
    ....:        plt.gray()
    ....:        plt.title(names[index])
    ....:        plt.subplot(2, 4, index+5)
    ....:        plt.imshow(output)
    ....:
In [49]: plt.show()
```

Observe that the results are of comparable quality to those obtained with the previous method:

Sharpening and blurring

There are many situations that produce blurred images:

- Incorrect focus at acquisition
- Movement of the imaging system
- The point-spread function of the imaging device (like in electron microscopy)
- Graphic-art effects

For blurring images, we could replicate the effect of a point-spread function by performing convolution of the image with the corresponding kernel. The Gaussian filter that we used for denoising performs blurring in this fashion. In the general case, convolution with a given kernel can be done with the routine `convolve` from the module `scipy.ndimage`. For instance, for a constant kernel supported on a 10 x 10 square, we could do as follows:

```
In [50]: from scipy.ndimage import convolve; \
    ....: from skimage.data import page
In [51]: kernel = np.ones((10, 10))/100.; \
    ....: blurred = convolve(page(), kernel)
```

To emulate the blurring produced by movement too, we could convolve with a kernel as created here:

```
In [52]: from skimage.draw import polygon
In [53]: x_coords = np.array([14, 14, 24, 26, 24, 18, 18]); \
    ....: y_coords = np.array([ 2, 18, 26, 24, 22, 18,  2]); \
    ....: kernel_2 = np.zeros((32, 32)); \
    ....: kernel_2[polygon(x_coords, y_coords)]= 1.; \
    ....: kernel_2 /= kernel_2.sum()
In [54]: blurred_motion = convolve(page(), kernel_2)
```

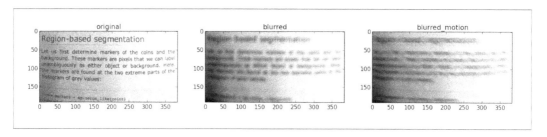

In order to reverse the effects of convolution when we have knowledge of the degradation process, we perform deconvolution. For example, if we have knowledge of the point-spread function, in the module `skimage.restoration` we have implementations of the Wiener filter (`wiener`), the unsupervised Wiener filter (`unsupervised_wiener`), and Lucy-Richardson deconvolution (`richardson_lucy`).

We perform deconvolution on blurred images by applying a Wiener filter. Note the huge improvement in readability of the text:

```
In [55]: from skimage.restoration import wiener
In [56]: deconv = wiener(blurred, kernel, balance=0.025, clip=False)
```

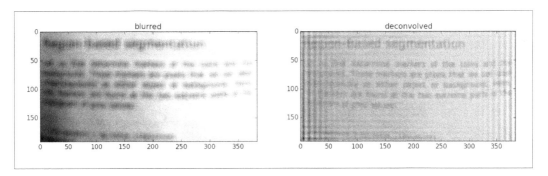

Inpainting

We define inpainting as the replacement of lost or corrupted parts of the image data (mainly small regions or to remove small defects).

There are ongoing efforts to include an implementation of Crimini's algorithm for inpainting in skimage. Until that day comes, there are two other options outside of the SciPy stack—an implementation of Alexandru Telea's Fast Marching Method and an implementation based on fluid dynamics (in particular, the Navier-Stokes equation). These two implementations can be called from the routine `inpaint` in the `imgproc` module of OpenCV. We use Telea's algorithm to illustrate the power of this technique: consider as a test image the checkerboard `skimage.data.checkerboard` to which we have removed a region.

```
In [57]: from skimage.data import checkerboard
In [58]: image = checkerboard(); \
   ....: image[25:100, 25:75] = 0.
In [59]: mask = np.zeros_like(image); \
   ....: mask[25:100, 25:75] = 1.
```

```
In [60]: from cv2 import inpaint, INPAINT_TELEA, INPAINT_NS
In [61]: inpainted = inpaint(image, mask, 1, INPAINT_TELEA)
```

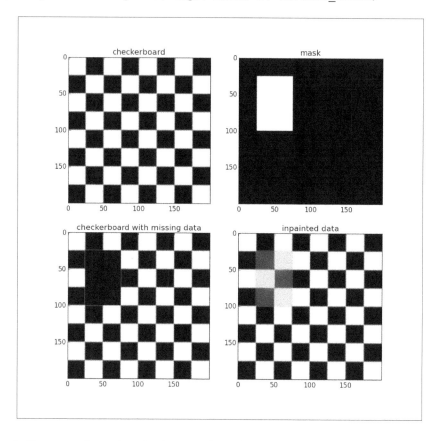

The result illustrates how inpainting works—the intensity of the pixels is obtained from nearby known values. It is thus no surprise that the inpainted region, yet preserving the geometry of the image, does not compute the correct set of intensities, but the most logical instead.

For more information on the image-processing module `imgproc` of OpenCV-Python, follow the API reference at `http://docs.opencv.org/modules/imgproc/doc/imgproc.html`.

Inpainting is very useful for removing unwanted objects from pictures. Observe, for example, in the page image `skimage.data.page`, the effect of removing large areas containing line breaks, and inpainting them with the Navier-Stokes algorithm:

```
In [62]: image = page(); \
    ....: image[36:46, :] = image[140:, :] = 0
In [63]: mask = np.zeros_like(image); \
    ....: mask[36:46, :] = mask[140:, :] = 1
In [64]: inpainted = inpaint(image, mask, 5, INPAINT_NS)
```

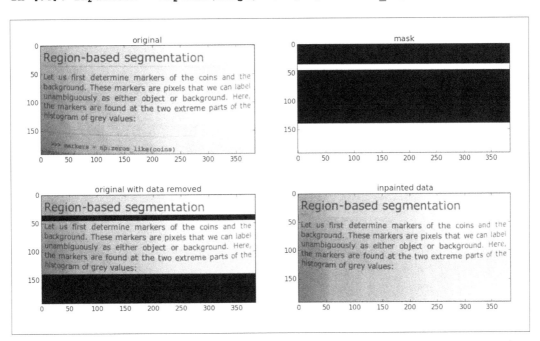

Image analysis

The aim of this section is the extraction of information from images. We are going to focus on two cases:

- Image structure
- Object recognition

Image structure

The goal is the representation of the contents of an image using simple structures. We focus on one case alone: **image segmentation**. We encourage the reader to explore other settings, such as quadtree decompositions.

Segmentation is a method to represent an image by partition into multiple objects (segments); each of them sharing some common property.

In the case of binary images, we can accomplish this by a process of labeling, as we have shown in a previous section. Let's revisit that technique with an artificial image composed by 30 random disks placed on a 64 x 64 canvas:

```
In [1]: import numpy as np, matplotlib.pyplot as plt
In [2]: from skimage.draw import circle
In [3]: image = np.zeros((64, 64)).astype('bool')
In [4]: for k in range(30):
   ...:     x0, y0 = np.random.randint(64, size=(2))
   ...:     image[circle(x0, y0, 3)] = True
   ...:
In [5]: from scipy.ndimage import label
In [6]: labels, num_features = label(image)
```

The variable labels can be regarded as another image, where each of the different objects found in the original image have been given a different number. The background of the image is also considered one more object, and received the number 0 as a label. Its visual representation (on the right in the following figure) presents all the objects from the image, each of them with a different color:

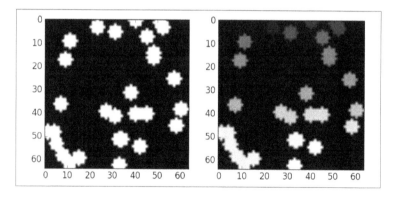

For gray-scale or color images, the process of segmentation is more complex. We can often reduce such an image to a binary representation of the relevant areas (by some kind of thresholding operation after treatment with morphology), and then apply a labeling process. But this is not always possible. Take, for example, the coins image `skimage.data.coins`. In this image, the background shares the same range of intensity as many of the coins in the background. A thresholding operation will result in failure to segment effectively.

We have more advanced options:

- Clustering methods using, as distance between pixels, the difference between their intensities/colors
- Compression-based methods
- Histogram-based methods, where we use the peaks and valleys in the histogram on an image to break it into segments
- Region-growing methods
- Methods based on partial differential equations
- Variational methods
- Graph-partitioning methods
- Watershed methods

From the standpoint of the SciPy stack, we have mainly two options, a combination of tools from `scipy.ndimage` and a few segmentation routines in the module `skimage.segmentation`.

> There is also a very robust set of implementations via bindings to the powerful library **Insight Segmentation and Registration Toolkit** (ITK). For general information on this library, the best resource is its official site `http://www.itk.org/`.
>
> We use a simplified wrapper build on top of it: The Python distribution of `SimpleITK`. This package brings most of the functionality of ITK through bindings to Python functions. For documentation, downloads, and installation, go to `http://www.simpleitk.org/`.
>
> Unfortunately, at the time this book is being written, the installation is very tricky. Successful installations depend heavily on your Python installation, computer system, libraries installed, and more.

Let's see, by example, the usage of some of these techniques on the particularly tricky image `skimage.data.coins`. For instance, to perform a simple histogram-based segmentation, we could proceed along these lines:

```
In [8]: from skimage.data import coins; \
   ...: from scipy.ndimage import gaussian_filter
In [9]: image = gaussian_filter(coins(), sigma=0.5)
In [10]: plt.hist(image.flatten(), bins=128); \
   ....: plt.show()
```

 Note how we first performed a smoothing of the original image by convolution with a spherical Gaussian filter. This is standard procedure to cancel possible unwanted signal and obtain cleaner results.

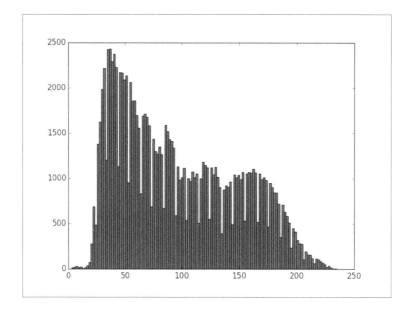

There seems to be a clear valley around the intensity `80`, in between peaks at intensities `35` and `85`. There is another valley around the intensity `112`, followed by a peak at intensity `123`. There is one more valley around intensity `137`, followed by a last peak at intensity `160`. We use this information to create four segments:

```
In [11]: level_1 = coins()<=80; \
   ....: level_2 = (coins()>80) * (coins()<=112); \
   ....: level_3 = (coins()>112) * (coins()<=137); \
   ....: level_4 = coins()>137
```

```
In [12]: plt.figure(); \
    ....: plt.subplot2grid((2,4), (0,0), colspan=2, rowspan=2); \
    ....: plt.imshow(coins()); \
    ....: plt.gray(); \
    ....: plt.subplot2grid((2,4),(0,2)); \
    ....: plt.imshow(level_1); \
    ....: plt.axis('off'); \
    ....: plt.subplot2grid((2,4),(0,3)); \
    ....: plt.imshow(level_2); \
    ....: plt.axis('off'); \
    ....: plt.subplot2grid((2,4), (1,2)); \
    ....: plt.imshow(level_3); \
    ....: plt.axis('off'); \
    ....: plt.subplot2grid((2,4), (1,3)); \
    ....: plt.imshow(level_4); \
    ....: plt.axis('off'); \
    ....: plt.show()
```

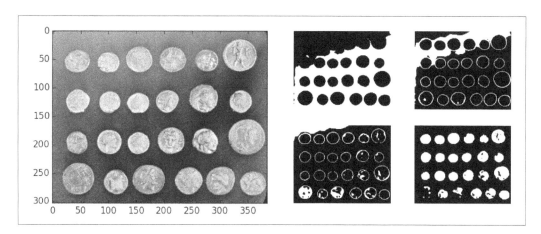

With a slight modification of the fourth level, we shall obtain a decent segmentation:

```
In [13]: from scipy.ndimage.morphology import binary_fill_holes
In [14]: level_4 = binary_fill_holes(level_4)
In [15]: labels, num_features = label(level_4)
```

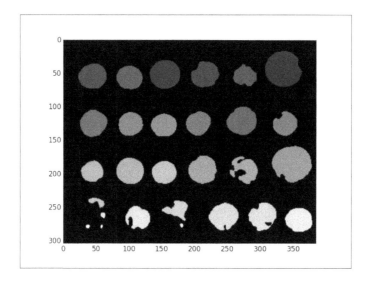

The result is not optimal. The process has missed a good segmentation of some coins in the fifth column, but mainly in the lowest row.

Improvements can be made if we provide a marker for each of the segments we are interested in obtaining. For instance, we can assume that we know at least one point inside of the locations of those 24 coins. We could then use a watershed transform for that purpose. In the module `scipy.ndimage`, we have an implementation of this process based on the iterative forest transform:

```
In [17]: from scipy.ndimage import watershed_ift
In [18]: markers_x = [50, 125, 200, 255]; \
    ....: markers_y = [50, 100, 150, 225, 285, 350]
In [19]: markers = np.zeros_like(image).astype('int16'); \
    ....: markers_index = [[x,y] for x in markers_x for y in
    ....:                   markers_y]
In [20]: for index, location in enumerate(markers_index):
    ....:     markers[location[0], location[1]] = index+5
    ....:
In [21]: segments = watershed_ift(image, markers)
```

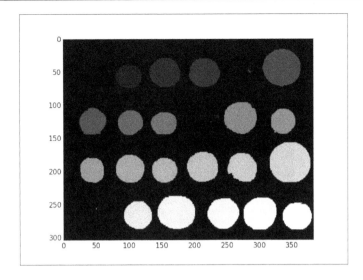

Not all coins have been correctly segmented, but those that had been corrected are better defined. To further improve the results of watermarking, we could very well work on the accuracy of the markers or include more than just one point for each desired segment. Note what happens when we add just one more marker per coin, to better segment three of the missing coins:

```
In [23]: markers[53, 273] = 9; \
    ....: markers[130, 212] = 14; \
    ....: markers[270, 42] = 23
In [24]: segments = watershed_ift(image, markers)
```

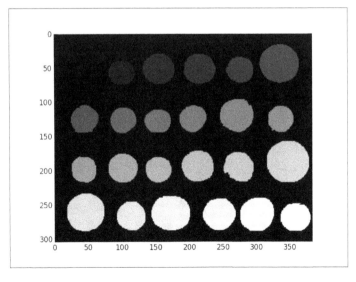

We can further improve the segmentation by employing, for instance, a graph-partitioning method, like a random walker:

```
In [26]: from skimage.segmentation import random_walker
In [27]: segments = random_walker(image, markers)
```

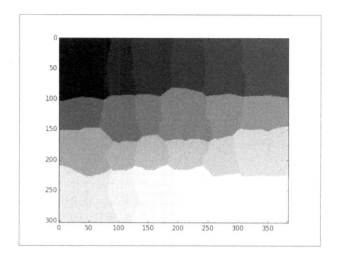

This process correctly breaks the image in to 24 well-differentiated regions, but does not resolve well the background. To take care of this situation, we manually mark, with a `-1` — those regions we believe are background. We can use the previously calculated masks `level_1` and `level_2` — they clearly represent the image background for these purposes:

```
In [29]: markers[level_1] = markers[level_2] = -1
In [30]: segments = random_walker(image, markers)
```

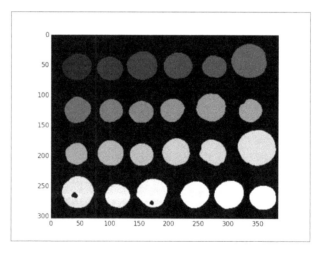

For other segmentation techniques, browse the different routines in the module `skimage.segmentation`.

Object recognition

Many possibilities arise. Given an image, we might need to collect the location of simple geometric features like edges, corners, linear, circular or elliptical shapes, polygonal shapes, blobs, and so on. We might also need to find more complex objects like faces, numbers, letters, planes, tanks, and so on. Let's examine some examples that we can easily code from within the SciPy stack.

Edge detection

An implementation of Canny's edge detector can be found in the module `skimage.feature`. This implementation performs a smoothing of the input image, followed by vertical and horizontal Sobel operators, as an aid for the extraction of edges:

```
In [32]: from skimage.feature import canny

In [33]: edges = canny(coins(), sigma=3.5)
```

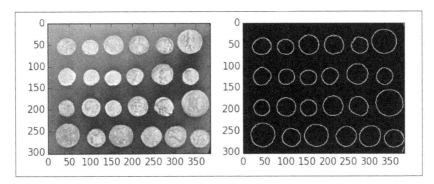

Line, circle, and ellipse detection

For detection of these basic geometric shapes, we have the aid of the Hough transform. Robust implementations can be found both in the module `skimage.transform`, and in the `imgproc` module of OpenCV. Let's examine the usage of the routines in the former by tracking these objects for an artificial binary image. Let's place an ellipse with center (10, 10) and radii 9 and 5 (parallel to the coordinate axes), a circle with center (30, 35) and radius 8, and a line between the points (0, 3) and (64, 40):

```
In [35]: from skimage.draw import line, ellipse_perimeter, \
    ....:                         circle_perimeter
```

```
In [36]: image = np.zeros((64, 64)).astype('bool'); \
   ....: image[ellipse_perimeter(10, 10, 9, 5)] = True; \
   ....: image[circle_perimeter(30, 35, 15)] = True; \
   ....: image[line(0, 3, 63, 40)] = True
```

To use the Hough transform for a line, we compute the corresponding H-space (the accumulator), and extract the location of its peaks. In the case of the line version of the Hough transform, the axes of the accumulator represent the angle `theta` and distance from the origin `r` in the Hesse normal form of a line $r = x \cos(\theta) + y \sin(\theta)$. The peaks in the accumulator then indicate the presence of the most relevant lines of the given image:

```
In [37]: from skimage.transform import hough_line, hough_line_peaks
In [38]: Hspace, thetas, distances = hough_line(image); \
   ....: hough_line_peaks(Hspace, thetas, distances)
Out[38]:
(array([52], dtype=uint64),
 array([-0.51774851]),
 array([ 3.51933702]))
```

This output means there is only one significant peak in the H-space of the Hough transform of the image. This peak corresponds to a line with the Hesse angle `-0.51774851` radians, `3.51933702` units from the origin:

$$3.51933702 = \cos(-0.51774851)x + \sin(-0.51774851)y$$

Let's see the original image together with the detected line:

```
In [39]: def hesse_line(theta, distance, thickness):
   ....:     return lambda i, j: np.abs(distance - np.cos(theta)*j \
   ....:                              - np.sin(theta)*i) < thickness
   ....:
In [40]: peak, theta, r = hough_line_peaks(Hspace, thetas, distances)
In [41]: detected_lines = np.fromfunction(hesse_line(theta, r, 1.),
   ....:                                  (64, 64))
```

 Note the inversion of the roles of the coordinates `i` and `j` in the definition of `hesse_line`. Why did we have to perform this artificial change of coordinates?

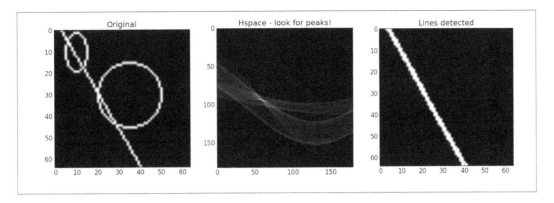

The detection of circles and ellipses obeys a similar philosophy of computing an accumulator in some H-space, and tracking its peaks. For example, if we are seeking circles with radii equal to `15`, and we wish to recover their centers, we could issue something along the following lines:

```
In [43]: from skimage.transform import hough_circle

In [44]: detected_circles = hough_circle(image,radius=np.array([15]))

In [45]: np.where(detected_circles == detected_circles.max())

Out[45]: (array([0]), array([30]), array([35]))
```

The array `detected_circles` has shape (`1, 64, 64`). The first index of the last output is thus irrelevant. The other two reported indices indicate that the center of the detected circle is precisely (`30, 35`).

Blob detection

We may regard a blob as a region of an image where all its pixels share a common property. For instance, after segmentation, each of the found segments is technically a blob.

There are some relevant routines in the module `skimage.feature` to this effect, `blob_doh` (a method based on determinants of Hessians), `blob_dog` (by differential of Gaussians), and `blob_log` (a method based on the Laplacian of Gaussians). The first approach ensures the extraction of more samples, and is faster than the other two:

```
In [46]: from skimage.data import hubble_deep_field; \
   ....: from skimage.feature import blob_doh; \
   ....: from skimage.color import rgb2gray

In [47]: image = rgb2gray(hubble_deep_field())

In [48]: blobs = blob_doh(image)
```

```
In [49]: plt.figure(); \
    ....: ax1 = plt.subplot(121); \
    ....: ax1.imshow(image); \
    ....: plt.gray(); \
    ....: ax2 = plt.subplot(122); \
    ....: ax2.imshow(np.zeros_like(image))
Out[49]: <matplotlib.image.AxesImage at 0x105356d10>
In [50]: for blob in blobs:
    ....:     y, x, r = blob
    ....:     c = plt.Circle((x, y),r,color='white',lw=1,fill=False)
    ....:     ax2.add_patch(c)
    ....:
In [51]: plt.show()
```

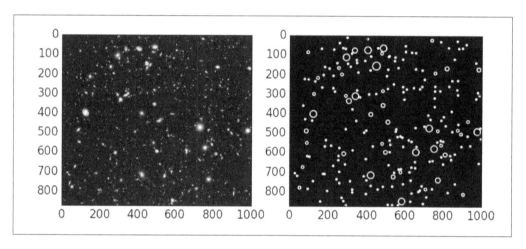

Corner detection

A corner is that location where two nonaligned edges intersect. This is one of the most useful operations in image analysis, since many complex structures require a careful location of these features. The applications range from complex object or motion recognition, to video tracking, 3D modeling, or image registration.

In the module `skimage.feature`, we have implementations of some of the best-known algorithms to solve this problem:

- **FAST corner detection** (features from accelerated segment test): `corner_fast`

- **Förstner corner detection** (for subpixel accuracy): `corner_foerstner`

- **Harris corner measure response** (the basic method): `corner_harris`

- **Kitchen and Rosenfeld corner measure response**: `corner_kitchen_rosenfeld`

- **Moravec corner measure response**: This is simple and fast, but not capable of detecting corners where the adjacent edges are not perfectly straight: `corner_moravec`

- **Kanade-Tomasi corner measure response**: `corner_shi_tomasi`

We also have some utilities to determine the orientation of the corners or their subpixel position.

Let's explore the occurrence of corners in `skimage.data.text`:

```
In [52]: from skimage.data import text; \
   ....: from skimage.feature import corner_fast, corner_peaks, \
   ....:                             corner_orientations
In [53]: mask = np.ones((5,5))
In [54]: corner_response = corner_fast(text(), threshold=0.2); \
   ....: corner_pos = corner_peaks(corner_response); \
   ....: corner_orientation = corner_orientations(text(), corner_pos,
   ....:                                           mask)
In [55]: for k in range(5):
   ....:     y, x = corner_pos[k]
   ....:     angle = np.rad2deg(corner_orientation[k])
   ....:     print "Corner ({}, {}) orientation {}".format(x,y,angle)
   ....:
Corner (178, 26) orientation -146.091580713
Corner (257, 26) orientation -139.929752986
Corner (269, 30) orientation 13.8150253413
Corner (244, 32) orientation -116.248065313
Corner (50, 33) orientation -51.7098368078
In [56]: plt.figure(); \
```

```
    ....: ax = plt.subplot(111); \
    ....: ax.imshow(text()); \
    ....: plt.gray()
In [57]: for corner in corner_pos:
    ....:     y, x = corner
    ....:     c = plt.Circle((x, y), 2, lw=1, fill=False, color='red')
    ....:     ax.add_patch(c)
    ....:
In [58]: plt.show()
```

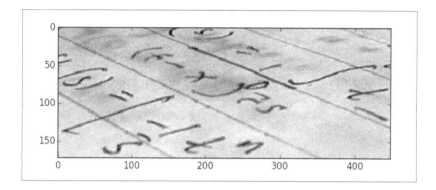

Beyond geometric entities

Object detection is not limited to geometric entities. In this subsection, we explore some methods of tracking more complex objects.

In the scope of binary images, a simple correlation is often enough to achieve a somewhat decent object recognition. The following example tracks most instances of the letter e on an image depicting the first paragraph of Don Quixote by Miguel de Cervantes. A tiff version of this image has been placed at https://github.com/blancosilva/Mastering-Scipy/tree/master/chapter9:

```
In [59]: from scipy.misc import imread
In [60]: quixote = imread('quixote.tiff'); \
    ....: bin_quixote = (quixote[:,:,0]<50); \
    ....: letter_e = quixote[10:29, 250:265]; \
    ....: bin_e = bin_quixote[10:29, 250:265]
In [61]: from scipy.ndimage.morphology import binary_hit_or_miss
In [62]: x, y = np.where(binary_hit_or_miss(bin_quixote, bin_e))
In [63]: plt.figure(); \
```

```
    ....: ax = plt.subplot(111); \
    ....: ax.imshow(bin_quixote)
Out[63]: <matplotlib.image.AxesImage at 0x113dd8750>
In [64]: for loc in zip(y, x):
    ....:     c = plt.Circle((loc[0], loc[1]), 15, fill=False)
    ....:     ax.add_patch(c)
    ....:
In [65]: plt.show()
```

```
  0
        1   In a village of La Mancha, the name of which I have no desire to call to mind,
        2   there lived not long since one of those gentlemen that keep a lance in the
        3   lance-rack, and old buckler, a lean hack, and a greyhound for coursing.  An olla
100     4   of rather more beef than mutton, a salad on most nights, scraps on Saturdays,
        5   lentils on Fridays, and a pigeon or so extra on Sundays, made away with
        6   three-quarters of his income.  The rest of it went in a doublet of fine cloth
200     7   and velvet breeches and shoes to match for holidays, while on week-days he made
        8   a brave figure in his best homespun.  He had in his house a housekeeper past
        9   forty, a niece under twenty, and a lad for the field and market-place, who used
300    10   to saddle the hack as well as handle the bill-hook.  The age of this gentleman
       11   of ours was bordering on fifty; he was of a hardy habit, spare, gaunt-featured,
       12   a very early riser and a great sportsman.  They will have it his surname was
       13   Quixada or Quesada (for here there is some difference of opinion among the
400    14   authors who write on the subject), although from reasonable conjectures it
       15   seems plain that he has called Quexana.  This, however, is of but little
       16   importance to our tale; it will be enough not to stray a hair's breadth from
500    17   the truth in the telling of it.
       18
      0        200       400       600       800      1000      1200
```

Small imperfections or a slight change of size in the rendering of the text makes correlation an imperfect detection mechanism.

An improvement that can be applied to gray-scale or color images is through the routine matchTemplate in the module imgproc of OpenCV:

```
In [66]: from cv2 import matchTemplate, TM_SQDIFF
In [67]: detection = matchTemplate(quixote, letter_e, TM_SQDIFF); \
    ....: x, y = np.where(detection <= detection.mean()/8.)
In [68]: plt.figure(); \
    ....: ax = plt.subplot(111); \
    ....: ax.imshow(quixote)
Out[68: <matplotlib.image.AxesImage at 0x26c7da890>]
In [69]: for loc in zip(y, x):
```

```
    ....:        r = pltRectangle((loc[0], loc[1]), 15, 19, fill=False)
    ....:        ax.add_patch(r)
    ....:
In [70]: plt.show()
```

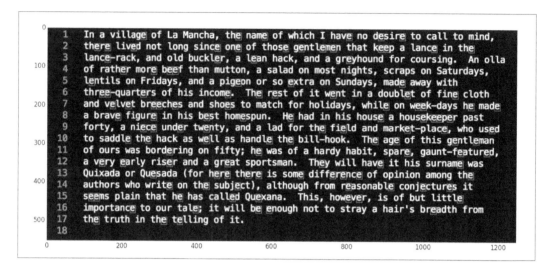

All the letter e's have been correctly detected now.

Let's finish the chapter with a more complex case of object detection. We are going to employ a Haar feature-based cascade classifier: this is an algorithm that applies a machine learning based approach to detect faces and eyes from some training data.

First, locate in your OpenCV installation folder the subfolder haarcascades. In my Anaconda installation, for instance, this is at /anaconda/pkgs/opencv-2.4.9-np19py27_0/share/OpenCV/haarcascades. From that folder, we are going to need the databases for frontal faces (haarcascade_frontalface_default.xml), and eyes (haarcascade_eye.xml):

```
In [71]: from cv2 import CascadeClassifier; \
    ....: from skimage.data import lena
In [72]: face_cascade = CascadeClassifier('haarcascade_frontalface_
default.xml'); \
    ....: eye_cascade = CascadeClassifier('haarcascade_eye.xml')
In [73]: faces = face_cascade.detectMultiScale(lena()); \
    ....: eyes = eye_cascade.detectMultiScale(lena())
In [74]: print faces
```

```
[[212 199 179 179]]
In [75]: print eyes
[[243 239  53  53]
 [310 247  40  40]]
```

The result is the detection of one face and two eyes. Let's put this all together visually:

```
In [76]: plt.figure(); \
   ....: ax = plt.subplot(111); \
   ....: ax.imshow(lena())
Out[76]: <matplotlib.image.AxesImage at 0x269fabed0>
In [77]: x, y, w, ell = faces[0]; \
   ....: r = plt.Rectangle((x, y), w, ell, lw=1, fill=False); \
   ....: ax.add_patch(r)
Out[77]: <matplotlib.patches.Rectangle at 0x26a837cd0>
In [78]: for eye in eyes:
   ....:     x,y,w, ell = eye
   ....:     r = plt.Rectangle((x,y),w,ell,lw=1,fill=False)
   ....:     ax.add_patch(r)
   ....:
In [79]: plt.show()
```

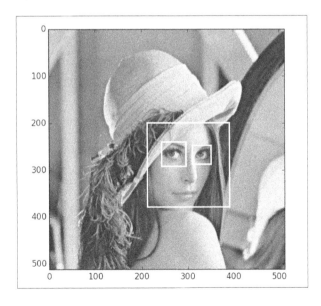

Summary

In this chapter, we have seen how the SciPy stack helps us solve many problems in Mathematical Imaging, from the effective representation of digital images, to their efficient storage, compression, and processing, modifying, restoring, or analyzing them. Although certainly exhaustive, this chapter scratches but the surface of this challenging field of engineering. One could easily write another 400 pages just devoted to this subject, and I invite you to further study the possibilities of the module `scipy.ndimage`, the imaging toolkit `skimage`, and the Python bindings for the libraries OpenCV or SimpleITK.

The chapter also closes my vision of what mastering the SciPy stack means. In truth, this vision only focused on the relational aspect between the scientific applications and the mathematical theory behind the needed routines. No efforts have been put forth to tackle the techniques of speeding up the codes by binding with other languages, for example. While this is an interesting and relevant topic, I defer to other more technical monographs on computer science for that aspect.

Index

lossy compression schemes, image compression 334-336

M

machine geometry 211
machine learning 219, 283
magnetic resonance imaging (MRI) device 5
Markov Chain Monte Carlo (MCMC) method 278
Mathematical Imaging
  about 311
  image acquisition 311
  image analysis 312
  image compression 312
  image editing 312
Mathematical morphology 321-323
matrices
  constructing, in matrix class 18
  constructing, in ndarray class 14-17
  creating 14
matrix addition 29, 30
matrix class
  matrices, constructing in 18
matrix equations
  about 40
  back and forward substitution 41
  banded matrices 41-44
  generic square matrices 44-50
  least squares 51
  regularized least squares 53
matrix equation solvers 53
matrix factorizations, based on eigenvalues
  about 54
  Schur decomposition 58, 59
  spectral decomposition 54-57
matrix factorizations, related to solving matrix equations
  about 38
  relevant factorizations 38
matrix functions
  about 35, 36
  computing 37, 38
Matrix Market Exchange format 5, 21
matrix multiplication 29, 30

MeanShift 299, 300
methods, for inputting sparse matrices
  Block Sparse Row 19
  Compressed Sparse Column 19
  Compressed Sparse Row 19
  Coordinate 19
  data, indices, and pointers 21
  data, rows, and columns 21
  diagonal storage 19
  dictionary of keys 19, 20
  fancy indexing 20
  row-based linked list 19
Mini-batch Kmeans 308
MINPACK
  URL 66
Monte-Carlo simulation 153
Moore-Penrose pseudo-inverse 32
mpmath
  about 168
  URL 168
multivariate calculus 324-326
multivariate interpolation 68, 81-91

N

Naive Bayes 290
ndarray class
  matrices, constructing in 14-17
nearest neighbors 205-207, 291, 292
nearest-neighbors interpolation 69
networkx
  about 23
  URL 23
Newton-Cotes quadratures 115
Newton methods 154-158
noise, cases
  Gaussian noise 349
  Poisson noise 349
  salt and pepper 349
  Speckle 349
noise reduction 349-353
nonlinear approximation 66
non-linear equations 129, 130
nonlinear least squares approximation 95-102
non-symmetric banded square matrix 43

QR decomposition 39
QR factorization 52
QUADPACK libraries, netlib repositories
  URL 117
quantitative variables
  relationship between 251

R

range 208
range searching 208, 209
real generalized Schur decomposition 58
real Schur decomposition 58
regression
  about 256
  ensemble methods 264
  linear regression beyond ordinary
    least-squares 263
  ordinary least-squares regression for
    large datasets 262
  ordinary linear regression for
    moderate-sized datasets 256-262
  support vector machines 263
regularized least squares 53
relevant factorizations
  about 38
  Cholesky decomposition 39
  pivoted LU decomposition 38
  QR decomposition 39
  singular value decomposition 39
run-length encoding 332
Rutherford-Boeing Exchange format 5

S

salt and pepper noise 349
scalar multiplication 29, 30
scatterplots
  and correlation 252-256
Schur decomposition
  about 58, 59
  complex generalized Schur
    decomposition 58
  complex Schur decomposition 58
  real generalized Schur decomposition 58
  real Schur decomposition 58
scikit-image library
  URL 312

scikit-learn
  URL 219
scipy.cluster
  URL 218
scipy.misc
  URL 108
scipy.misc library 314
scipy.ndimage library
  URL 312
scipy.sparse.csgraph module
  about 199
  URL 200
scipy.sparse module 199
scipy.special
  URL 129
scipy.stats
  URL 218
scipy.stats.mstats
  URL 218
segment 180
segmentation 358
sharpening 353-355
shortest paths 188, 199-202
Sigmoid correction 346
simple iterative solvers 139-141
SimpleITK
  URL 359
singular value decomposition 39, 52, 53
skimage.color module
  URL 318
sparse matrices
  constructing 19-26
  reference link 4
Sparse Matrix Collection
  URL 56
Speckle 349
spectral clustering 309
spectral decomposition 54-56
spectral embedding 296
Spline interpolation 77-80
static problems, combinatorial
    computational geometry
  about 179
  convex hulls 187-191
  shortest paths 188, 199-202
  triangulations 187, 195-199
  Voronoi diagrams 187, 192-195

statical filters 326
statistical inference
  about 219, 275
  estimation 275
  estimation of parameters 277
  hypothesis testing 275
  interval estimation 275, 282
Steiner points 196
stochastic gradient descent (SGD) 262
stochastic methods 153-155
SuperLU
  URL 13
support vector classification 286-288
support vector machines 263
swirl
  performing 337, 338
symbolic differentiation 107-109
symbolic integration 110-113
symbolic setting, probability 231-234
symbolic solution of differential
      equations 166-168
SymPy
  URL 168
sympy libraries
  about 130
  URL 130
sympy.stats
  URL 218
systems of nonlinear equations
  about 135-138
  broyden method 141, 142
  large scale solvers 145
  Powell's hybrid solver 142-145
  simple iterative solvers 139-141

T

Theano
  URL 108
time plots 246
time series
  analysis 264-274
traces 30, 31
transformations, of domain
  about 336
  geometric transformations 338-341

rescale 337
resize 337
swirl, performing 337, 338
transition matrix 4
transposes 31, 32
triangle 183
triangulations 187, 195-199
two-dimensional geometric shapes,
      skimage.draw module
  circles 314
  ellipses 314
  lines 314
  polygons 314

U

unconstrained optimization for
      multivariate functions
  about 150-152
  conjugate gradient methods 154-160
  Newton methods 154-158
  stochastic methods 153-155
unconstrained optimization for univariate
      functions 146-148
univariate interpolation
  about 68, 69
  Hermite interpolation 73, 74
  Lagrange interpolation 70-72
  nearest-neighbors interpolation 69
  piecewise polynomial interpolation 75-77
  Spline interpolation 77-80

V

VBGMM 304
Voronoi diagrams 187, 192-195

W

wavelet decompositions
  performing 329, 330

Z

zero spline 78

Thank you for buying
Mastering SciPy

About Packt Publishing

Packt, pronounced 'packed', published its first book, *Mastering phpMyAdmin for Effective MySQL Management*, in April 2004, and subsequently continued to specialize in publishing highly focused books on specific technologies and solutions.

Our books and publications share the experiences of your fellow IT professionals in adapting and customizing today's systems, applications, and frameworks. Our solution-based books give you the knowledge and power to customize the software and technologies you're using to get the job done. Packt books are more specific and less general than the IT books you have seen in the past. Our unique business model allows us to bring you more focused information, giving you more of what you need to know, and less of what you don't.

Packt is a modern yet unique publishing company that focuses on producing quality, cutting-edge books for communities of developers, administrators, and newbies alike. For more information, please visit our website at www.packtpub.com.

About Packt Open Source

In 2010, Packt launched two new brands, Packt Open Source and Packt Enterprise, in order to continue its focus on specialization. This book is part of the Packt Open Source brand, home to books published on software built around open source licenses, and offering information to anybody from advanced developers to budding web designers. The Open Source brand also runs Packt's Open Source Royalty Scheme, by which Packt gives a royalty to each open source project about whose software a book is sold.

Writing for Packt

We welcome all inquiries from people who are interested in authoring. Book proposals should be sent to author@packtpub.com. If your book idea is still at an early stage and you would like to discuss it first before writing a formal book proposal, then please contact us; one of our commissioning editors will get in touch with you.

We're not just looking for published authors; if you have strong technical skills but no writing experience, our experienced editors can help you develop a writing career, or simply get some additional reward for your expertise.

Learning SciPy for Numerical and Scientific Computing

ISBN: 978-1-78216-162-2   Paperback: 150 pages

A practical tutorial that guarantees fast, accurate, and easy-to-code solutions to your numerical and scientific computing problems with the power of SciPy and Python

1. Perform complex operations with large matrices, including eigenvalue problems, matrix decompositions, or solution to large systems of equations.

2. Step-by-step examples to easily implement statistical analysis and data mining that rivals in performance any of the costly specialized software suites.

NumPy Cookbook

ISBN: 978-1-84951-892-5   Paperback: 226 pages

Over 70 interesting recipes for learning the Python open source mathematical library, NumPy

1. Do high performance calculations with clean and efficient NumPy code.

2. Analyze large sets of data with statistical functions.

3. Execute complex linear algebra and mathematical computations.

Please check **www.PacktPub.com** for information on our titles

NumPy Beginner's Guide

Second Edition

ISBN: 978-1-78216-608-5   Paperback: 310 pages

An action packed guide using real world examples of the easy to use, high performance, free open source NumPy mathematical library

1. Perform high performance calculations with clean and efficient NumPy code.

2. Analyze large data sets with statistical functions.

3. Execute complex linear algebra and mathematical computations.

IPython Interactive Computing and Visualization Cookbook

ISBN: 978-1-78328-481-8   Paperback: 512 pages

Over 100 hands-on recipes to sharpen your skills in high-performance numerical computing and data science with Python

1. Leverage the new features of the IPython notebook for interactive web-based big data analysis and visualization.

2. Become an expert in high-performance computing and visualization for data analysis and scientific modeling.

3. A comprehensive coverage of scientific computing through many hands-on, example-driven recipes with detailed, step-by-step explanations.

Please check **www.PacktPub.com** for information on our titles

www.ingramcontent.com/pod-product-compliance
Lightning Source LLC
LaVergne TN
LVHW081330050326
832903LV00024B/1095